Genetic Engineering and its Applications

Genetic Engineering and its Applications

Dr. Arvind Kumar

RANDOM PUBLICATIONS
NEW DELHI (INDIA)

Genetic Engineering and its Applications

ISBN 978-93-5111-591-5

Published in 2015 in India by

Reprint 2019

RANDOM PUBLICATIONS

4376-A/4B, Gali Murari Lal, Ansari Road
New Delhi-110 002
Phone : +9111-43580356, 011-23289044, 011-43142548
e-mail: sales@randompublications.com,
info@randompublications.com, randomexports@gmail.com

Type Setting by : Friends Media, Delhi-110089
Printed at : Mehra Printers, Delhi-110 092

Preface

Students learn how engineers apply their understanding of DNA to manipulate specific genes to produce desired traits, and how engineers have used this practice to address current problems facing humanity. They learn what genetic engineering means and examples of its applications, as well as moral and ethical problems related to its implementation. Students fill out a flow chart to list the methods to modify genes to create GMOs and example applications of bacteria, plant and animal GMOs. Genetic engineers have developed genetic recombination techniques to manipulate gene sequences in plants, animals and other organisms to express specific traits. Applications for genetic engineering are increasing as engineers and scientists work together to identify the locations and functions of specific genes in the DNA sequence of various organisms. Once each gene is classified, engineers develop ways to alter them to create organisms that provide benefits such as cows that produce larger volumes of meat, fuel- and plastics-generating bacteria, and pest-resistant crops.

I would like to thank my team for standing beside me throughout my career and writing this book. My special thanks go to "Random Publications" who have published the book.

– Dr. Arvind Kumar

Contents

1

Genes: Definition and Structure

HISTORICAL DEFINITION OF A GENE

The observable characteristics of an organism are referred to collectively as its phenotype. Humans have known for thousands of years that organisms within species – cattle, grain crops, dogs, date palms, horses and, of course, humans – differ from one another phenotypically and that phenotypic variations are, to a large extent, heritable. Long before the mechanisms of heredity were understood in any depth, humans were selectively breeding plants and domestic animals to enhance desirable qualities and to eliminate undesirable ones. Early theories of heredity proposed that offspring were a concoction of fluids derived from one or both parents and that inherited characteristics were somehow determined by the properties of these fluids. Biologists up to the time of Charles Darwin, including Darwin himself, believed that inherited characteristics were literally dissolved like sugar in water. This notion led to a difficult problem: Characteristics conveyed from one generation to the next in the form of liquids would become more and more dilute with the passage of time. How could natural selection produce evolutionary change if favourable variations in phenotype were as evanescent as a drop of juice?

When genetics was a young science most of us considered the genes as the determinants for the "unit-characters", each gene responsible for a separate quality of an organism. The facts of Mendelian inheritance showed us that there are very small hereditary differences—differences between a contrasting pair of characters.

And it seemed obvious to assume that in such a case the genes themselves occur in pairs, one member of the pair causing white flower-colour and one blue colour, one causing waxy endosperm in corn and one horny. It is very curious that this idea should have taken such a firm grip upon the imagination of the earliest geneticists. If we compare genetics with other sciences—with physics, chemistry, astronomy—we generally see that it is not self-evident to assume that where we see a difference, or where we see two contrasting qualities, two different contrasting causes occur. We speak of hot and cold, of light and dark, of wet and dry. But we do not really think of a separate cause for wetness and a separate one for dryness—a relative presence or absence of water does very well as an explanation—and the same is true in all the other examples. To assume the existence of pairs of genes in each pair of which the resultant action would be contrasting is an unnecessary complication.

The alternative is a much simpler hypothesis which was proposed by Bateson and by myself, and which went under the name of the "presence- and-absence "theory of the gene. We assumed that in a monogenetic difference—the smallest "Mendelizing" difference—we were dealing with the absence or presence of just one gene.

At first this hypothesis met with a veritable storm of protests. To some of my colleagues it seemed the height of folly that the absence of anything should ever determine an inheritable quality. On the face of it, this objection seems very strange; for it seems quite logical to speak of silence as determined by a purely negative absence of noise, of dryness determined by the absence of water. But of course the apparent absurdity of thinking of the absence of a gene as causing an inheritable quality grew out of Weismann's conception of determinants, inherited protoplasmic entities that should be each responsible for just one particular property of a living being.

The idea, now generally accepted by biologists, that the qualities of an organism develop as a result of growth and development, this development being caused by a co-operation of thousands of factors, some of these factors being environmental, while some others—the genes— influence the course of development from within. If we compare a plant having pale-pink flowers with one having dark-pink ones, we assume that the development which leads to the production of the plant with the pale flowers is due to the co-operation of thousands of causes and substances. But in the other plant one additional substance is present, whose action at the appropriate developmental stage allows a further transformation of the pink colour into a darker one. It is therefore not the absence of this gene which directly causes pink colour; it is the presence of the thousand-and-one genes which both plants have in common.

If we knock off a man's hat we may see that he is red-headed. He does not become red-headed as a result of the absence of the hat, but we can see the result of all the other things which cause his hair to be red, provided we take away that hat. If in the course of this story we substitute an inherited gene which helps to transform red hair into black, the reasoning remains essentially the same.

Very soon after the original presence-absence hypothesis was first published it was found that in its original shape this hypothesis was not adequate to explain all the facts. It was found that very often in inheritance we are not dealing with just two contrasting qualities, but with three or more alternatives. In snapdragons we have a series of varieties that gradually grade into each other, starting from yellow through different intermediate stages to red. If we call those forms A, B, C, we find that any pair of them between them show a difference in just one gene. When we mate A with B, we will get only A and B plants in following generations; if we combine B with Z, no 's or C"s, but only B's and Z>'s will be found in the second generation.

The best explanation of such series of multiple alleles is this: that in such cases we are dealing not with the presence and complete absence of one gene, but with different quantities of one and the same gene in the one spot on the chromosome where our gene is located. Instead of the presence-absence hypothesis, we now prefer to speak of a hypothesis about quantitative differences in inheritance.

It is highly probable that in certain instances we are really concerned with presence versus complete absence; in other cases, however, intermediate stages may occur in which fewer molecules of the gene are present per cell-nucleus.

This quantitative gene-theory is closely bound up with the hypothesis and which is now generally accepted by geneticists—namely, that the genes are not living protoplasma globules, but that they are non-living chemical substances, that have only this in common: that they are each the ferment for their own synthesis. With Alexander, we may speak of reproductive catalysis. We assume that each molecule of a gene, between every time the cells, in growing tissue, divide, attracts and helps to combine from the ambient building stones just one new molecule that will in every particular be its exact duplicate.

The genes are combined longitudinally in long series—the chromosomes—and it is the chromosomes whose movements and combinations give rise to the phenomena we notice as Mendelian segregation and redistribution of genes. The number of the chromosomes is highly constant for every plant species; and sometimes, as in wheat and barley, different sections of the group are distinguished by different numbers of chromosomes. In that case the numbers are always multiples

Many genes of different kinds of a common basis number, and it is very probable that in such a series the higher chromosome numbers may have originated by reduplication of entire chromosome sets during the course of evolution of the group. In the cells of the plants those chromosomes, made up of genes, occur in pairs.

The homologous chromosomes of each pair are not necessarily identical. We know that in a plant heterozygous in respect to one gene, the chromosomes of the corresponding pair are different in that particular spot ("locus "). The two chromosomes of a homologous pair have originally been derived from the two parents, one from the pollen-cell and one from the ovule.

The same is true of all the other pairs. The somatic (body) cells have a double set of chromosomes; the germ-cells have only a single set—one of each kind. In ordinary growing tissue every chromosome doubles its mass (and probably the quantity of every gene present on it) between two cell divisions.

The two new chromosomes each consisting of part of the two original ones are identical, the break occurring on each at exactly the same spot. And when cell-division commences, the chromosomes are drawn apart to different poles of the cell in such a way that each daughter cell obtains a complete set of chromosomes that is identical with the original set.

The exception is that a "reduction division" occurs when germ-cells are formed. In this case there is a second cell division without a preceding chromosomal doubling. At this reduction division the two chromosomes of a homologous pair become separated one from the other. The result is that in a plant, heterozygous in respect to one gene, Ay in which all the somatic cells are Aa, the germ-cells will be either A or a or only half the germ-cells contain the full quantity of gene A.

When the break does not happen to occur at exactly the same "locus" in the two original chromosomes, one of the new chromosomes may have one gene in duplicate and the other one may be lacking it. Mutations may arise in this way from small irregularities during crossing-over. for this reason, be of four kinds: ABy Ab, aB and ab.

To a certain extent this redistribution of distinct genes over different germ-cells results in a redistribution of the qualities which were shown by the parents of the heterozygous plant. If we cross a plant with AB to an ab plant, the hybrid will be AaBb and as this hybrid will be making four kinds of germ-cells, there will be all sorts of offspring, with different combinations of the qualities in which the grandparents differed.

We were just speaking of genes located in different chromosomal pairs. But sometimes plants are heterozygous in respect to two genes that happen to be located in the same chromosome. In this case those chromosomes tend to be transmitted as such, in their entirety, so that the distribution of the two genes is not mutually independent. Such genes are "coupled ".

All kinds, all degrees of coupling have been met with, from complete or almost complete coupling to very loose coupling. Morgan proposed a hypothesis which is now very well proven and established—namely, that the degree of coupling is dependent on the greater or lesser degree of contiguity of the genes. Genes that lie close together are closely coupled, those that are far apart must lie farther apart on the same chromosome.

The mechanism which is thought to account for coupling is the phenomenon of "crossing-over ". Sometimes chromosomal strands seem to touch at one or more places, and when the strands break apart, the top half of one strand may be attached to the bottom part of the other.

A new pair of chromosomal strands originates, but if on both sides of a break there are loci where the strands differ in chromosomal content, those loci are redistributed, so that, as far as the independent inheritance of separate genes is concerned, crossing-over may help to redistribute the genes, even if they are located in the same chromosome.

GENOTYPIC VARIATION

How can students use Fast Plants to investigate the contribution of the genotype to the phenotype? Fast Plants, rapid cycling Brassica rapa, are genotypically variable in that they have a genetically controlled mating system that prevents self-fertilization and favours out-crossing among individuals. As a consequence, even seed stocks selected for uniformity of specific phenotypes and genotypes exhibit considerable variation for other traits.

Fig. Rosette, dwarf, basic, elongated internode

The Wisconsin Fast Plants® Programme has developed a number of genetic stocks of rapid cycling Brassica rapa for genetic investigations on the nature and inheritance of variation.

Some WFP stocks contain distinctive mutant phenotypes, *e.g.*, anthocyaninless plant, anl, yellow green plant, ygr1, rosette, ros, and male sterile, mst2, that are suitable for Mendelian genetics.

Other stocks exhibit phenotypes whose expression may vary continually and which may be quantified as discrete or countable units, *e.g.*, number of hairs, or as estimates of size, *e.g.* petite dwarf, dwf1, or of intensity of colour saturation, *e.g.*, purple anthocyanin. These quantitative phenotypes may be conditioned by a few or many genes and normally require numerical descriptions in which the statistical notations of population size, (n), range (r), arithmetic mean (x), and standard deviation (s) are applied. Yet other stocks have been developed to combine both simply inherited mutant genotypes and quantitatively expressed phenotypes. An important part of WFP is the continuing development and improvement of seed stocks for uses in genetics.

ACTION OF PLANT HORMONES

Plant hormones are at the heart of many plant developmental processes. They determine critical growth parameters, such as the rate of cell division or elongation, as well as developmental decisions, for instance where along a primary root a lateral root should be formed.Different plant hormones control different developmental processes to varying degrees, sometimes in the same, sometimes in opposite directions. Whether and how the overlapping influence of different hormones on traits is coordinated is largely unknown. Part of the research in our lab aims to determine how the major plant hormone auxin, arguably the most important in shaping the plant, regulates downstream events that lead to discrete morphological change. After all, those changes require structural rearrangements on the cellular and organ level after hormone signaling. One tool to answer this question are the *hy5* and *hyh* mutants of Arabidopsis, which display constitutively increased auxin signaling.

Another important question is the nature of crosstalk between hormone pathways, that is the impact of more than one hormone signaling pathway on the same trait or molecular event.

Does such crosstalk simply reflect a quantitative add up of different hormone signals at the point of convergence (In the way the water flow rate through a river is determined by the contribution of its two tributaries downstream of their junction), or is there real crosstalk happening upstream (The two tributaries that will meet have built-in dams, and if so, do the dam gatekeepers communicate with each other to intentionally regulate the water flow beyond the river junction)?

Much to our surprise, we isolated an important regulator of hormone pathway interaction in the root through our natural variation project: loss of function of the *BRX* gene results in reduced root growth because *BRX* is required for the biosynthesis of the hormone brassinolide, whose concentration is in turn rate-limiting for auxin action. Since *BRX* is itself under auxin control, this creates a feedback loop required for optimal root growth. Whether an equivalent loop works in the shoot, and whether this feedback is conserved in other species are some of the questions we are currently interested in with regard to this project, generously funded by the Swiss National Science Foundation.

Variation

If you have grown Fast Plants you will have noticed variation among the individual plants within the group. Variation can range from a little to a lot. Variation is one of the fundamental characteristics of life. All organisms exhibit some variation among individuals. Understanding the ways that variation is manifested in organisms, how it comes to be expressed through the development of the individual, and how it is transmitted from one individual to the next generation of individuals are central themes in biology.

Working with Fast Plants will enrich a student's understanding of variation. By observing the growth and development of a Fast Plants through the various stages in the life cycle, students will become aware of many visible features, or phenotypes, that make up the organism. Only, however, upon close observation of a population of plants, will they become aware that the characteristics observed on one plant vary more or less on other plants. Such is the nature of variation.

PHENOTYPIC VARIATION

Phenotypic variation, *e.g.* plant height at a particular stage of development, is considered to be the expression of the genetic makeup (genotype) of the individual as it interacts with the environment. Variation in plant height among individuals in a population is therefore due to variation in the interaction between the genotype and the environment, as expressed through the development of each individual plant.

In order to be useful in an experiment the phenotype must be described using terms that are widely understood and easily communicated.

For these reasons scientists have agreed upon various standards or descriptors to describe characteristics in the natural world. The choice of how to describe what you observe is important, because it will determine the kinds of descriptors used and establish the basis for recording, analyzing and communicating results.

ENVIRONMENTAL VARIATION

Much can be learned about the role of light, temperature, and nutrients on plant development from experiments in which one or more environmental parameters are changed. Because Fast Plants are highly responsive to changes in the environment, they are ideal for examining the role of the environment on the expression of phenotypic variation.

Although Fast Plants are able to grow within a wide range of environmental conditions, for most investigations it is recommended that they be grown under uniform and ideal conditions.

In this way variation arising from sub-optimal conditions of environment will be minimized.

The Wisconsin Fast Plants Information Document (WFPID) Understanding the Environment describes how to provide and maintain the various physical, chemical and biotic components of the environment that are most suitable for Fast Plants.

GENETIC VARIATION

Our main interest is to isolate genes that are responsible for the intra- and interspecific morphological variation in plants. Therefore, the analysis of natural genetic variation is the starting point for the majority of individual projects in our lab.

For the isolation of genes involved in morphological variation, exploiting natural genetic variation offers some distinct advantages as compared to classical mutagenesis experiments. Most importantly, it allows the identification of naturally occurring, differentially active gene alleles that are likely targets for the evolution of morphological variation. Because naturally occurring lines do not carry major deleterious mutations that impair their survival in the wild, the approach also counter-selects against the isolation of mutant alleles that result in strong phenotypes as commonly detected in mutagenesis experiments.

This applies in particular to wild populations with a high level of inbreeding, such as in our favourite model organism, *Arabidopsis thaliana*.

So why has this approach not been followed earlier? Mainly because the isolation of genes that modify quantitative traits has technically not been feasible. However, the rich biological and genomic resources, such as the fully sequenced genome, have turned Arabidopsis into an ideal model system to investigate the molecular basis of natural genetic variation at the gene level.

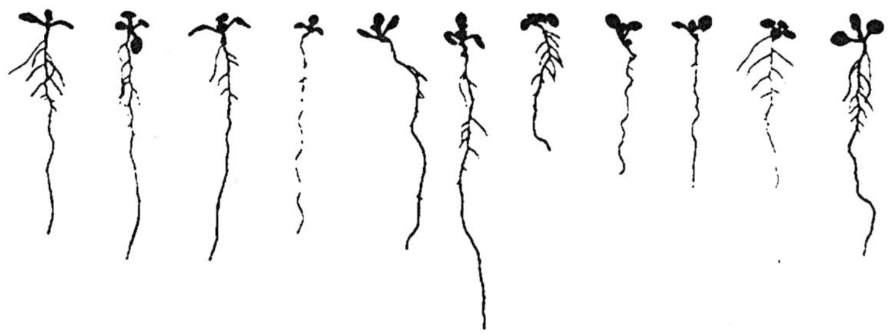

Fig. Natural morphological variation of root system architecture between wild isolates of Arabidopsis. Most of this variation results from the influence of genetic factors.

Plant organs are formed in a continuous, post-embryonic manner by ordered cell divisions and expansions. The extent of growth ultimately determines organ shape. Thus, we are primarily interested in genes that modulate growth rate. To isolate such genes, we focus on the root system.

This is because of our interest in root biology per se, but also because roots grow from a so-called meristematic region at their tip, which produces concentric tissue layers through controlled cell proliferation, elongation and differentiation. This feature enables us to reduce the problem of accurately measuring growth from three dimensions to a single one.

Nevertheless, analysis of root system growth requires care and highly reproducible growth conditions, because root system development is very plastic and responds to many macro- and micro-environmental cues. The natural variation in root system morphology between Arabidopsis accessions collected from different locations and grown in identical growth chamber conditions on tissue culture media.So far, we have isolated an important regulator of root system architecture from the accession Uk-1. This gene is a major so-called quantitative trait locus (QTL), which we gave the name *BREVIS RADIX (BRX)*, latin for "short root". Other efforts in this area include attempts to isolate two other QTLs for root growth located in other accessions.

2

Genetic Engineering

Genetic engineering, also called genetic modification, is the direct manipulation of an organism's genome using biotechnology. New DNA may be inserted in the host genome by first isolating and copying the genetic material of interest using molecular cloning methods to generate a DNA sequence, or by synthesizing the DNA, and then inserting this construct into the host organism. Genes may be removed, or "knocked out", using a nuclease. Gene targeting is a different technique that uses homologous recombination to change an endogenous gene, and can be used to delete a gene, remove exons, add a gene, or introduce point mutations. An organism that is generated through genetic engineering is considered to be a genetically modified organism (GMO). The first GMOs were bacteria in 1973 and GM mice were generated in 1974. Insulin-producing bacteria were commercialized in 1982 and genetically modified food has been sold since 1994. Glofish, the first GMO designed as a pet, was first sold in the United States December in 2003. Genetic engineering techniques have been applied in numerous fields including research, agriculture, industrial biotechnology, and medicine. Enzymes used in laundry detergent and medicines such as insulin and human growth hormone are now manufactured in GM cells, experimental GM cell lines and GM animals such as mice or zebrafish are being used for research purposes, and genetically modified crops have been commercialized.

Genetically modified plants are created by the process of genetic engineering, which allows scientists to move genetic material between organisms with the aim of changing their characteristics. All organisms are composed of cells that contain the DNA molecule. Molecules of DNA form units of genetic information, known as genes. Each organism has a genetic blueprint made up of DNA that determines the regulatory functions of its cells and thus the characteristics that make it unique.

Prior to genetic engineering, the exchange of DNA material was possible only between individual organisms of the same species. With the advent of genetic engineering in 1972, scientists have been able to identify specific genes associated with desirable traits in one organism and transfer those genes across species boundaries into another organism. A gene from bacteria, virus, or animal may be transferred into plants to produce genetically modified plants having changed characteristics. Thus, this method allows mixing of the genetic material among species that cannot otherwise breed naturally. The success of a genetically improved plant depends on the ability to grow single modified cells into whole plants. Some plants like potato and tomato grow easily from single cell or plant tissue. Others such as corn, soy bean, and wheat are more difficult to grow.

After years of research, plant specialists have been able to apply their knowledge of genetics to improve various crops such as corn, potato, and cotton. They have to be careful to ensure that the basic characteristics of these new plants are the same as the traditional ones, except for the addition of the improved traits.

The world of biotechnology has always moved fast, and now it is moving even faster. More traits are emerging; more land than ever before is being planted with genetically modified varieties of an ever-expanding number of crops. Research efforts are being made to genetically modify most plants with a high economic value such as cereals, fruits, vegetables, and floriculture and horticulture species.

Biological Cell

It only takes one biological cell to create an organism. In fact, there are countless species of single celled organisms, and indeed multi-cellular organisms like ourselves.

A single cell is able to keep itself functional by owning a series of '*miniature machines*' known as *organelles*. The following list looks at some of these organelles and other characteristics typical of a fully functioning cell.

- *Mitochondrion:* An important cell organelle involved in respiration
- *Cytoplasm:* A fluid surrounding the contents of a cell and forms a vacuole
- *Golgi Apparatus:* The processing area for the creation of a glycoprotein
- *Endoplasmic Reticulum:* An important organelle heavily involved in protein synthesis.
- *Vesicles: Packages* of substances that are to be used in the cell or secreted by it.
- *Nucleus*: The "brain" of a cell containing genetic information that determines every natural process within an organism.
- *Cell Membrane:* Also known as a plasma membrane, this outer layer of a cell assists in the movement of molecules in and out the cell plays both a structural and protective role
- *Lysosomes:* Membranous sacs that contain digestive enzymes

Cell Wall

A structure that characteristically is found in plants and prokaryotes and not animals that plays a structural and protective role.

Cell Specialisation

Cells can become specialised to perform a particular function within an organism, usually as part of a larger tissue consisting of many of the same cells working in tandem:

- *Nerve cells* to operate as part of the nervous system to send messages back and forth via the brain at the centre of the nerve system.
- Skin cells for waterproof protection and protection against pathogens in the open air environment.
- *Xylem* tubes to transport water around plants and to provide structural support for the plant as a whole.

Cells combine their efforts in these tissue types to perform a common cause. The task of the specialised cell will determine in what way it is going to be specialised, because different cells are suited to different purposes, as illustrated in the above list:

- Muscle cells are long and smooth in structure and their elastic nature allows these cells to perform flexible movements, just as they do in our own body's.
- Some *white blood cells* contain powerful digestive enzymes to eliminate pathogens by breaking them down to the molecular level.
- Cells at the back of the eye are sensitive to light stimuli, and thus can *interpret* differences in light intensity which can in turn be interpreted by our nervous system and brain.

Many of these cells contain organelles, though after some cells are specialised, they do not possess particular characteristics as they do not require them to be there. *i.e.* efficiency is the key, no resources are wasted and the resources available are put to their idyllic optimum.

The Cell Membrane

The *cell membrane*, otherwise known as the plasma membrane is a semi-permeable structure consisting mainly of *phospholipid* (fat) molecules and proteins.

They are structured in a *fluid mosaic model*, where a double layer of phospholipid molecules provide a barrier accompanied by proteins.

It is present round the circumference of a cell to acts as a barrier, keeping foreign entities out the cell and its contents (like cytoplasm) firmly inside the cell.

The plasma membrane allows only selected materials to pass in and out of a cell, and is thus known as a selectively permeable membrane.

Cell Transport

There are three methods in which ions are transported through the cell membrane into the cell,

- *Active Transport*: Active transport is the transport of molecules with the active assistance of a carrier that can transport the material against a natural *concentration gradient*.
- *Passive Transport (Diffusion):* The movement of molecules from areas of high concentration (*i.e.* outside a cell) to areas of low concentration (*i.e.* within a cell) via a carrier. This process does not require energy.
- *Simple Diffusion:* The movement of molecules from areas of high concentration to areas of low concentration in a free state. *Osmosis* of water involves this type of diffusion through a selectively permeable membrane (*i.e.* plasma membrane)

The Breakdown of Materials in a Cell

In cells, sometimes it is required to breakdown more complex molecules into more simple molecules, which can then be 're-built' into what is needed by the body with these new raw materials.

'*Pinocytosis*' where to contents of a structure (such as bacteria) are *drank*, essentially by breaking down molecules into a drinkable form.

'*Phagocytosis*' where contents are 'eaten'.

Absorption and Secretion

Absorption is the uptake of materials from a cells' external environment. Secretion is the ejection of material.

CELL RESPIRATION

As mentioned in the previous page on ATP, the process of *respiration* is split into 3 distinct areas that occur at different parts of the cell. Respiration involves the *oxidation* of foodstuff (*i.e.* glucose) in order to create ATP.

Respiration can occur with or without oxygen, *aerobic* and *anaerobic* respiration respectively.

Glycolysis

Glycolysis occurs in the *cytoplasm* of a cell where a 6 carbon glucose molecule (the broken down food that you ate earlier) is broken down by enzymes into a 3 carbon *pyruvic acid*.

The execution of this process requires 2 ATP, and produces a net gain of 2 ATP. The enzymes involved remove hydrogen from the glucose (oxidation) where they take these hydrogen atoms to the cytochrome system, explained soon. In anaerobic respiration, this is where the process ends, glucose is split into 2 molecules of pyruvic acid. When oxygen is present, pyruvic is broken down into other carbon compounds in the Kreb's Cycle. When it is not present, the pyruvic acid is broken down into lactic acid (or carbon dioxide and ethanol).

The Kreb's Cycle

When oxygen is present, respiration can harness more ATP from a single unit of glucose. The *pyruvic acid* from the *glycolysis* stage diffuses into a cell organelle called a mitochondrion (pl. *mitochondria*). These mitochondria are sausage shaped structures that host a large surface area for the respiration to occur on.

The pyruvic acid is then subject to more enzymes which break it down into a 2 carbon compound, as seen below. The diagram illustrates the *Kreb's cycle*, consisting of three main actions

- The carbon element is in an infinite cycle where the 2 carbon compound derived from pyruvic acid binds with the 4 carbon compound that is always present in the cycle.
- CO_2 is released, where the oxygen that is present in *aerobic respiration* combines with carbon from the carbon compounds which is released as CO_2. Hence the need for animals to breath out and expel this CO_2.
- Enzymes oxidize the carbon compounds and transport the hydrogen atoms to the cytochrome system.

The Cytochrome System

The cytochrome system, also known as the hydrogen carrier system (or the *electron transport system*) are where the reduced hydrogen carriers transport hydrogen atoms from the glycolysis and Kreb's cycle stages. The cytochrome system is found in the many *cristae* of mitochondria, which are tiny stalked particles found on its outer layer.

The system contains many 'hydrogen acceptors' which hydrogen can be added to. By following the path of a hydrogen atom, we can see how the cytochrome system works:

- Some *coenzymes* from earlier stages are transferred to the next coenzymes.
- B is then oxidised, therefore the coenzyme releases the hydrogen and energy is made available.
- The released hydrogen atom binds with 2 oxygen atoms (oxygen is available in aerobic respiration) which produces water, a by-product of respiration.

The diagram illustrates this flow of hydrogen within the cytochrome system and how energy is made available by the flow of these atoms. The green circles illustrate where energy is made available via oxidation. Overall their is a gain of 38 ATP from one molecule of glucose in aerobic respiration. The food that we eat provides glucose required in respiration. In plants, energy is also acquired via respiration, but the mechanism of delivering glucose to the respiration process is a little different.

DNA STRUCTURE AND DNA REPLICATION

The energy required by these cells and how energy is created in order for the cell to survive.The structure, type and functions of a cell are all determined by *chromosomes* that are found in the *nucleus* of a cell. These chromosomes are composed of DNA, the acronym for deoxyribonucleic acid.

This DNA determines all the characteristics of an organism, and contains all the genetic material that makes us who we are. This information is passed on from generation to generation in a species so that the information within them can be passed on for the offspring to harness in their lifetime.

Structure of DNA and Nucleotides

DNA is arranged into a *double helix* structure where spirals of DNA are intertwined with one another continuously bending in on itself but never getting closer or further away.

The following diagram illustrates a *nucleotide*, the building blocks of DNA

There are four different types of nucleotide possible in a DNA sequence, adenine, cytosine, guanine and thymine (can be replaced with A, C, G and T). There are billions of these nucleotides in our genome, and with all the possible permutations; this is what makes us unique. Nucleotides are situated in adjacent pairs in the double helix nature mentioned. The following rules apply in regards to what nucleotides pair with one another.

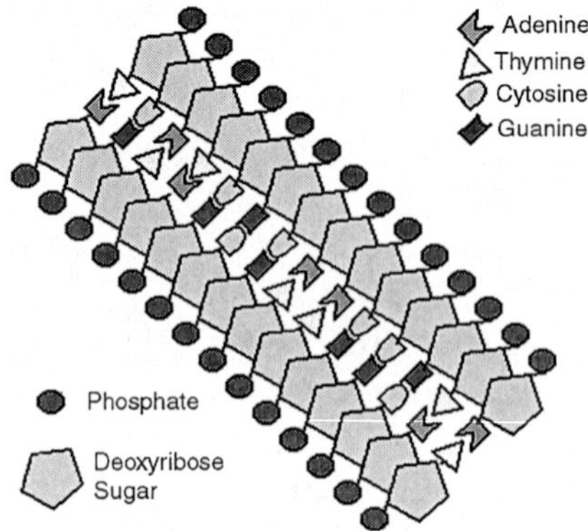

- There are four possible types of nucleotide, adenine, cytosine, guanine and thymine.

- Thymine and adenine can only make up a base pair
- Guanine and cytosine can only make up a base pair
- Therefore, thymine and cytosine would NOT make up a base pair, as is the case with adenine and guanine.

This is illustrated in the below diagram, using correct pairings of nucleotides

The diagram is two dimensional, remember that DNA is structured in a double helix fashion, as shown to the above right. This continuous sequence, and the sequence they are in determine an organisms' structural, physical and anatomical features.

DNA Replication

Cells do not live forever, and in light of this, they must pass their genetic information on to new cells, and be able to replicate the DNA to be passed on to offspring. It is also required that fragments of DNA (genes) have to be copied to code for particular bodily function.

It is essential that the replication of it is EXACT. In order for replication to occur, the following must be available

- The actual DNA to act as an exact template
- A pool of relevant and freely available nucleotides
- A supply of the relevant enzymes to stimulate reaction
- ATP to provide energy for these reactions

When replicating, the double helix structure uncoils so that each strand of DNA can be exposed. When they uncoil, the nucleotides are exposed so that the freely available nucleotides can pair up with them.

When all nucleotides are paired up with their new partners, they re-coil into the double helix. As there are two strands of DNA involved in replication, the first double helix produces 2 copies of itself via each strand.

It is said that the replicated DNA is semi-conservative, because it possesses 50percent of the original genetic material from its parent. These 2 new copies have the exact DNA that was in the previous one. This template technique allows genetic information to be passed from cell to cell and from parents to offspring.

PROTEIN SYNTHESIS

If you have jumped straight to this page, you may wish to look at the previous page about DNA, which gives background information on protein synthesis.As mentioned, a string of *nucleotides* represent the genetic information that makes us unique and the blueprint of who and what we are, and how we operate. Part of this genetic information is devoted to the synthesis of *proteins*, which are essential to our body and used in a variety of ways. Proteins are created from templates of information in our DNA, illustrated below:

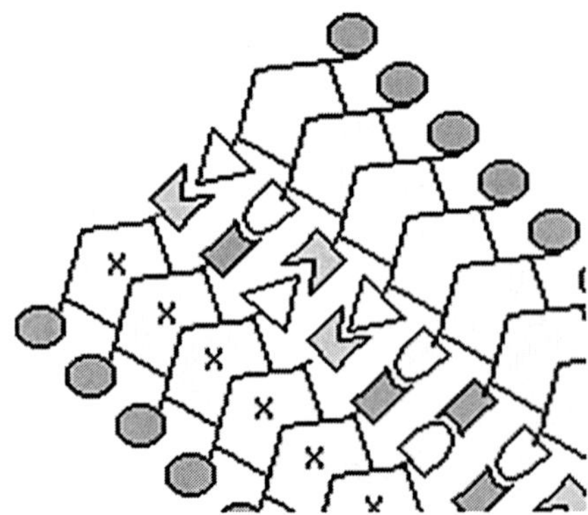

The X marked nucleotides are an example of a DNA sequence that would be used to code for a particular protein, with the sequence of these nucleotides determining which protein it is.

The sequence of these nucleotides are used to create amino acids, where chains of *amino acids* form to make a protein.

mRNA

This genetic information is found in the nucleus, though protein synthesis actually occurs in *ribosomes* found in the *cytoplasm* and on rough endoplasmic reticulum. If protein is to be synthesised, then the genetic information in the *nucleus* must be transferred to these ribosomes. This is done by *mRNA* (messenger ribonucleic acid). It is very similar to DNA, but fundamentally differs in two ways

- A base called *uracil replaces* all thymine bases in mRNA.
- The deoxyribose sugar in DNA in is *replaced* by ribose sugar in mRNA.

At the beginning of protein synthesis, just like DNA replication, the double helix structure of DNA uncoils in order for mRNA to replicate the genetic sequence responsible for the coding of a particular protein.

In the beginning, the DNA has uncoiled, allowing the mRNA to move in and transcribe (copy) the genetic information. If the code of DNA looks like this: G-G-C-A-T-T, then the mRNA would look like this C-C-G-U-A-A (remembering that uracil replaces thymine)

With the genetic information responsible for creating substances now available on the mRNA strand, the mRNA moves out of the nucleus and away from the DNA towards the ribosomes.

PROTEIN VARIETY

As mentioned in the previous two pages investigating protein synthesis, each consists of a successive chain of amino acids. The sequence of these amino acids determine which type of protein it is. It is synthesised from a DNA strand, each DNA strand involved in protein synthesis is responsible for producing a unique protein.

Types of Protein

Over time and diversity of organisms, a huge amount of proteins exist and perform a unique function in the body. Primarily, their are three types of protein

- *Fibrous Proteins*: These fibre like proteins are used for structural purposes in organisms. This is because fibrous proteins are arranged in long strands and are insoluble in water. Examples of use include providing a barrier in the cell wall of plants and myosin in skeletal muscle
- *Globular Proteins*: The polypeptide chains (protein chains) in globular proteins are folded together into a knot like shape essential in the fact that are present in the following
  1. *Enzymes*: Biological catalysts, enzymes are responsible speeding up reactions in an organism
  2. *Hormones*: Hormones are chemical messengers responsible for initialising a response in organisms. Some hormones have a regulatory effect,
  3. *Antibodies*: Antibodies are used to defend the body against foreign agents *e.g.* bacteria, fungi and viruses. The next page investigates these.
  4. *Structural Protein*: Globular proteins form part of the cell membrane, which has a structural role as well as a role in transporting ions in and out the cell.
- *Conjugated Proteins*: Conjugated proteins are essentially globular proteins that possess non-living substances, such as the haem found in haemoglobin, which possesses iron (a non-living substance)

Therefore proteins play a vital role in many of an organisms biological processes and their organs. The following page investigates cell defence against foreign agents, where proteins are playing their role in the form of antibodies.

BIOLOGICAL VIRUSES

The prime directive of all organisms is to reproduce and survive, which is also the case for viruses, which in most cases are considered a nuisance to humans.

Viruses

Viruses possess both living and non-living characteristics. The unique characteristic that differentiates viruses from other organisms is the fact that they require other organisms to host themselves in order to survive, hence they are deemed *obligate parasites*.

Viruses can be spread in the following exemplar ways:

- *Airborne*: Viruses that infect their hosts from the open air
- *Blood Borne*: Transmission of the virus between organisms when infected blood enters an organisms circulatory system

- *Contamination*: Caused from the consumption of materials by organisms such as water and food which have viruses within

Therefore viruses have many means of getting transmitted from one organism to another.

Cell Assimilation by a Virus

Viruses are tiny micro-organisms, and due to their size and simplicity, they are unable to replicate independently.

Therefore, when a virus is situated in a host, it requires the means to reproduce before it dies out without producing more viruses.

This is done by altering the genetic make up of a cell to start coding for materials required to make more viruses.

By altering the cell instructions, more viruses can be produced which in turn, can affect more cells and continue their existence as a species. The following is a step by step guide of how an example bacteriophage (a virus that infects bacteria) takes control of its host cell and reproduces itself.

- The virus approaches the bacteria and attaches itself to the cell membrane
- The tail gives the virus the means to thrust its genetic information into the bacteria
- Nucleotides from the host are 'stolen' in order for the virus to create copies of itself
- The viral DNA alters the genetic coding of the host cell to create protein coats for the newly create viral DNA strands
- The viral DNA enters its DNA coat
- The cell is swollen with many copies of the original virus and bursts, allowing the viruses to attach themselves to other nearby cells
- The process begins all over again with many more viruses attacking the hosts' cells

Without a means of defence, the host that is under attack from the virus would soon die.

Biological Cell Defence

Organisms must find a means of defence against antigens such a viruses described on the previous page. If this was not the case, bacteria, fungi and viruses would replicate out of control inside other organisms which would most likely already be extinct.

Therefore organisms employ many types of defence to stop this happening. Means of defence can be categorised into first and second lines of defence, with the first line usually having direct contact with the external environment.

First Lines of Defence

- *Skin* is an excellent line of defence because it provides an almost impenetrable biological barrier protecting the internal environment.
- *Lysozyme* is an enzyme found in tears and saliva that has powerful digestive capabilities, and can break down foreign agents to a harmless status before they enter the body.
- The clotting of blood near open wounds prevents an open space for antigens to easily enter the organism by coagulating the blood.
- *Mucus* and *cilia* found in the nose and throat can catch foreign agents entering these open cavities then sweep them outside via coughing, sneezing and vomiting.
- The *cell wall* of plants consists of fibrous proteins which provide a barrier to potential parasites (antigens).

If these first lines of defence fail, then there are further defences found within the body to ensure that the foreign agent is eliminated.

Second Lines of Defence

Second lines of defence deal with antigens that have bypassed the first lines of defence and still remain a threat to the infected organism.*Interferons* are a family of proteins that are released by a cell that is under attack by an antigen. These interferons attach themselves to *receptors* on the plasma membrane of other cells, effectively instructing it of the previous cells' situation.This tells these neighbouring cells that an antigen is nearby and instructs them to begin coding for antiviral proteins, which upon action, defend the cell by shutting it down. In light of this, any invading antigen will not be able to replicated its DNA (or mRNA) and *protein coat* inside the cell, effectively preventing the spread of it in the organism. These antiviral proteins provide the organism with protection against a wide range of viruses.

This action brought about by interferon is a defensive measure, while *white blood cells* in the second line of defence in animals can provide a means of attacking these antigens. One method of attacking antigens is by a method called *phagocytosis*, where the contents of the antigen are broken down by molecules called phagocytes. These phagocytes contain digestive enzymes in their lysosomes (an organelle in phagocytes) such as lysozyme. White blood cells such as a *neutrophil* or a monocyte are capable of undergoing phagocytosis, which is illustrated below.

- The bacterium inside the cell gives out chemical messages that are picked up by the phagocyte.
- The bacteria targets the cell as a possible host and moves towards it.
- The cell is prepared for this and the bacterium becomes trapped in a vacuole that forms around it.

- The bacterium is a sitting duck that is harmless at present.
- The lysosomes detect the bacterium and the digestive enzymes inside them begin to break the bacterium down.
- The remnants of the lysosome and bacterium materials are absorbed into the cytoplasm.

The above illustrates one method of ridding an organism of an internal threat caused by an antigen. This is a non-specific response to an antigen.

Passive and Active Types of Immunity

The previous page investigated the role of white blood cells in phagocytosis. *White blood cells* are also responsible for *antibody* formation. Certain antibodies are synthesised in response to the presence of certain *antigens*

Specific Immune Responses

Lymphocytes are a type of white blood cell capable of producing a *specific immune response* to unique antigens. Some of these lymphocytes are capable of entrapping antigens on their surface.When lymphocytes catch these antigens they can then begin to code for unique antibodies, structures that are capable of catching these antigens. The lymphocytes code for a particular antibody on response to a particular antigen. The antibody that is formed will be capable of catching free antigens therefore neutralising the threat as seen below.

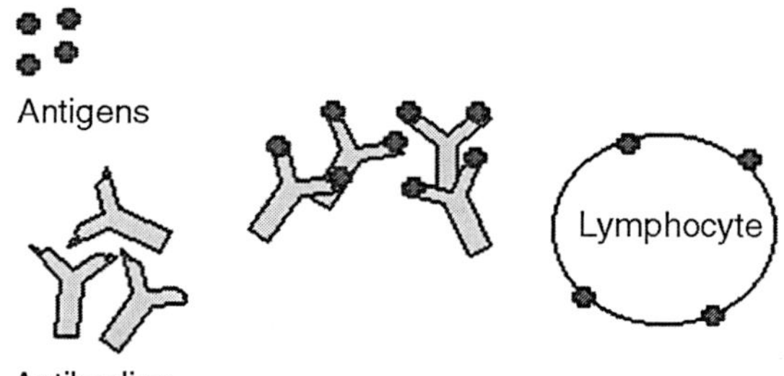

B lymphocytes (B Cells) produce free moving antibodies as above while *T lymphocytes* (T Cells) produce antibodies on their surface.

Types of Immunity

When attacked an organism has several means in which it can prepare to defend itself in event of attack.

- *Active Immunity*: *Vaccines* are used for health purposes to expose our bodies to a particular antigen. These antigens are usually killed or severely weakened to decrease their potency. After destroying these pathogens, the body stores some T cells as memory cells, due to the fact they code for a particular antigen and can be when needed. This memory in T cells can be a means of artificially acquiring immunity while a genuine attack by a pathogen is a naturally acquired type of immunity.

- *Passive Immunity*: This is where immunity to particular antigens as a result of genetic traits passed on from parents rendering the offspring immune to a particular pathogenic threat.

Plant Cell Defence

Hydrogen Peroxide

Plants release *hydrogen peroxide* in response to the presence of a fungal invasion, which attacks by piercing the cell wall of a plant and breaking it down. This hydrogen peroxide (chemical symbol H_2O_2) is a double edged sword in its defence against the antigen. *One Way*: Hydrogen peroxide stops the breakdown of the cell wall Certain pathogens will use pectinase, a digestive enzyme, to break down the cell wall barrier and invade the plant.

The *pectinase* released by the fungus must be stopped. H_2O_2 is involved in halting the action of this pectinase:

- The H_2O_2 is created and moves to the cell wall - the site of the invasion.
- It reacts in contact with and enzyme called *peroxidase*, which promotes the breakdown of *pectinase*.
- The foreign chemical is rendered useless.
- Threat of the cell wall being compromised is removed.

Another Way: Some of the H_2O_2 triggers the creation of phytoalexins

Phytoalexins are similar to the antiviral proteins previously mentioned as a secondary line of defence. Phytoalexins are a family of hormones that inhibit protein synthesis and thus "shut up shop" in the event of a pathogenic attack by halting the protein production process in our cells.

- Chemicals released by the fungus that are being used by it in its attack trigger a chemical response in the plasma membrane that makes the plant aware of the pathogens presence.
- Hydrogen peroxide from the plasma membrane triggers a chemical response to inform the nucleus of the infected cells of the current situation.
- mRNA from the nucleus is transported to ribosomes, as described in the protein synthesis section, where phytoalexins are to be produced (essentially protein synthesis coding for phytoalexins).
- The phytoalexins then take on a role similar to that of antiviral proteins, where the presence of a phytoalexin in a cell inhibits protein synthesis and therefore preventing growth of the foreign agent by removing all possible avenues of invasion for the pathogen, thus eliminating the threat.

Barriers Used by Plants in Defence

Lignin is a strong type of molecule that provides plants with a defensive structure similar to that of fibrous proteins. It acts as a barrier and can be found in wood and is characteristically found in plants that have recently endured pathogen attack.

Callose seals off sieve plates in the plant, effectively shutting off the transport of molecules around the organism. This is done to minimise the chance of the plant transporting infectious material around its own self, and halting the movement of materials that could be used by the pathogen in aid of replicating itself.

Ethylene promotes leaf abscission, and is done to sever the plant of dead or dying plant matter. This is done to prevent the spread of infected material, therefore sacrificing infected sections of plant is more economical than taking the risk of the infection spreading.

Galls and *tannins* are created by the plant to encapsulate foreign agents found within the plant. A gall is an instance where an infected cell becomes inflamed that contains tannins. These tannins play a protective role by segregating the foreign agent and its chemicals from the rest of the plant

All of the four previous pages have illustrated means of self defence against *pathogens* (fungi, viruses and bacteria).

Plant transformation

Genetic engineering of plants is much easier than that of animals. There are several reasons for this:

1. There is a natural transformation system for plants (the bacterium *Agrobacterium tumefaciens*),
2. Plant tissue can redifferentiate (a transformed piece of leaf may be regenerated to a whole plant), and
3. Plant transformation and regeneration are relatively easy for a variety of plants.

The soil bacterium *Agrobacterium tumefaciens* ("tumefaciens" meaning tumor-making) can infect wounded plant tissue, transferring a large plasmid, the Ti plasmid, to the plant cell. Part of the Ti (tumor-inducing) plasmid apparently randomly integrates into the chromosome of the plant.

The integrated part of the plasmid contains genes for the synthesis of

1. Food for the bacterium, and
2. *Plant hormones*: Genes from the Ti plasmid that are integrated in the plant chromosome are expressed at high levels in the plant.

Overproduction of the plant hormones leads to continuous growth of the transformed cells, causing plant tumors. Rapid, cancerous growth of the transformed plant tissue obviously is advantageous to the bacterium: more food gets produced.

The Ti plasmid has been genetically modified ("disarmed") by deleting the genes involved in the production of bacterial food and of plant hormones, and inserting a gene that can be used as a selectable marker. Selectable marker genes generally are coding for proteins involved in breakdown of antibiotics, such as kanamycin. Any gene of interest can be inserted into the Ti plasmid as well. In principle, one can thus transform any plant tissue, and select transformants by screening for antibiotic resistance.

However, unfortunately, there are some complications:

1. It has proven difficult to transform some monocots (grasses, etc.) by *Agrobacterium*, and
2. Regeneration of plants from tissue culture or leaf discs is not always possible.

A number of genetically engineered plant varieties have been developed. Traits that have been introduced by transformation include herbicide resistance, increased virus tolerance, or decreased sensitivity to insect or pathogen attack. Traditionally, most of such genetically engineered plants were tobacco, petunia, or similar species with a relatively limited agricultural application. However, during the past decade it now has become possible to transform major staples such as corn and rice and to regenerate them to a fertile plant. Increasingly, the transformation procedures used do not depend on *Agrobacterium tumefaciens*. Instead, DNA can be delivered into the cells by small, μm-sized tungsten or gold bullets coated with the DNA.

The bullets are fired from a device that works similar to a shotgun. The modernized device uses a sudden change in pressure of He gas to propel the particles, but the principle of "shooting" the DNA into the cell remains the same.

This DNA-delivery device is nicknamed "gene gun", and has been shown to work for DNA delivery into chloroplasts as well. Over the last several years, use of the "gene gun" has become a very common method to transform plants, and has been shown to be applicable to virtually all species investigated. Transformation of rice by this method is now routine. This is a very important development as rice is the most important crop in the world in terms of the number of people critically dependent on it for a major part of their diet.

Another method to get foreign genes into cereals is by electroporation: a jolt of electricity is used to puncture self-repairing holes in protoplasts (*i.e.*, the cell without the cell wall), and DNA can get in through these holes.

However, it is often very difficult to regenerate fertile plants from protoplasts of cereals. Non-etheless, significant advances in overcoming these practical difficulties have been made over the years.

Now even transgenic trees have been created: the gene for a coat protein of the plum pox virus has been introduced into apricot. The plum pox virus leads to the feared Sharka disease, for which there is no cure. The resulting transgenic tree shows a markedly decreased sensitivity to this virus. The reason why continuous exposure of the tree to the viral coat protein leads to tolerance against viral infection is not yet understood, however. Thus, now there are a number of different techniques to introduce foreign genes into plants. Essentially all major crop plants can be (and have been or are being) genetically engineered, the procedures are now routine and the frequency of success is very high. Even though genetically engineered crops are more costly than the usual ones, they have been rather readily accepted by US farmers provided that tangible benefits can be demonstrated. However, it is questionable whether the farmer in poorer countries can come up with the funds to "try out" and use the new crops.

Another issue in this respect is how genetically engineered crops are perceived by the consumer. Even though in the US there is little resistance to such crops as long as the products can be shown to be safe and advantageous, in other countries genetically modified foods are received poorly by the consumer.

It is unlikely that there is a rationally sound basis for this rather hostile reaction of the consumer, as most of the crops are the result of human manipulation and may have been treated with harmful herbicides and pesticides.

Time and education will need to be invested to provide consumers and consumer advocates with a balanced opinion on the acceptability of the origin of their foods. One area of particular concern for some people is the lack of labeling of genetically engineered foods, and legislation may be introduced to address this issue.

On the other hand, as so many plants (soybean, corn, etc.) are genetically modified and the nature of the genetic modification is not necessarily easy to explain, it may be simpler to label those foods that are guaranteed free of "genetically modified organisms" or their products. However, keep in mind that essentially all agricultural products have been genetically modified by traditional breeding, so it may be difficult to define what is actually free of genetically modified organisms.

BIOLOGICAL ENERGY - ADP AND ATP

ATP stands for Adenosine Tri-Phosphate, and is the energy used by an organism in its daily operations.

It consists of an *adenosine* molecule and three inorganic *phosphates*. After a simple reaction breaking down ATP to *ADP*, the energy released from the breaking of a molecular bond is the energy we use to keep ourselves alive.

ATP to ADP - Energy Release

This is done by a simple process, in which one of the phosphate molecules is broken off, therefore reducing the ATP from 3 phosphates to 2, forming ADP (Adenosine Diphosphate after removing one of the phosphates {Pi}). This is commonly wrote as ADP + Pi. When the bond connecting the phosphate is broken, *energy* is released.

While ATP is constantly being used up by the body in its biological processes, the energy supply can be bolstered by new sources of glucose being made available via eating food which is then broken down by the digestive system to smaller particles that can be utilised by the body.

On top of this, ADP is built back up into ATP so that it can be used again in its more energetic state. Although this conversion requires energy, the process produces a net gain in energy, meaning that more energy is available by re-using ADP+Pi back into ATP.

Glucose and ATP

Many ATP are needed every second by a cell, so ATP is created inside them due to the demand, and the fact that organisms like ourselves are made up of millions of cells. *Glucose*, a sugar that is delivered via the bloodstream, is the product of the food you eat, and this is the molecule that is used to create ATP. Sweet foods provide a rich source of readily available glucose while other foods provide the materials needed to create glucose.

This glucose is broken down in a series of *enzyme* controlled steps that allow the release of energy to be used by the organism. This process is called respiration.

Respiration and the Creation of ATP

ATP is created via respiration in both animals and plants. The difference with plants is the fact they attain their food from elsewhere.

In essence, materials are harnessed to create ATP for biological processes. The energy can be created via cell respiration. The process of respiration occurs in 3 steps (when oxygen is present):

- Glycolysis
- The Kreb's Cycle
- The Cytochrome System

APPLICATIONS OF PLANT GENETIC ENGINEERING

"Pharming" and "plantibodies"

An increasingly viable option is the production of highly valuable enzymes by plants and animals. In addition to production of human proteins in these organisms, other valuable proteins that are currently produced by microorganisms could very well be produced by higher organisms instead. Animal and plant "bioreactors" in some respects may be superior to recombinant bacterial systems, because eukaryotes glycosylate proteins. Whereas the glycosylation pattern may be species-specific, appropriate glycosylation is often required for protein function. Production through these organismal systems may also be cheaper than cell fermentation techniques. Two examples of production of human proteins in plants include the production of human serum albumin in transgenic tobacco and potato, and production of human insulin by tobacco.

In both cases, the produced protein appears to be fully effective in humans. Unfortunately, however, one cannot raise his/her insulin level by eating transgenic tobacco leaves, as the protein in most cases will be broken down to amino acids before it reaches the blood stream.

Therefore, in these cases one cannot escape the practice of protein isolation and purification before transgenic leaves are converted into drugs. Also antibodies are being produced in plants. Initially, the antibody's light and heavy chains were produced in different plants. But a subsequent cross of these two varieties resulted in progeny carrying assembled and functional antibodies.

Reversible male sterility in plants

As has been indicated earlier, heterozygous individuals often are healthier and stronger than homozygous ones. The only way to guarantee heterozygoiscity in plants is to make sure self-pollination cannot occur. For most crop plants it was very tedious or practically impossible to exclude selfing.

To exclude self-pollination, it would be good to introduce male sterility in plants: progeny from such plants are then expected to be 100percent heterozygous (assuming they were pollinated with pollen from an unrelated variety). To introduce male sterility, a promoter was identified that was turned on exclusively in tapetum cells (a tissue around the pollen sac that is essential for pollen production).

This promoter then was linked up to a gene coding for a bacterial ribonuclease (named barnase). This ribonuclease selectively chops up ribonucleic acids. The promoter/ribonuclease construct was then introduced into plants (canola, tobacco, you name it). Because the promoter allows expression only in tapetum cells, the gene construct disrupts only development of the tapetal tissue and its end product, pollen. Plants transformed with this construct were male-sterile but otherwise normal.

Although male-sterile plants are valuable for hybrid seed production, they have limited value when it comes to crop production. Fertility must be restored to crops such as wheat, rice, and tomato, in which the seed or fruit is the harvested product. Fortunately, the ribonuclease is inhibited very much by a simple protein, named barstar.

One can thus cross the male-sterile plant with a male-fertile variety in which the gene for barstar has been introduced, and the result is progeny with viable pollen and restored fertility.

A closely related approach has been criticized as "terminator technology" as it is seen by its critics as a way for companies to protect and enforce their patents.

Antisense RNA

Antisense RNA refers to nucleotide strands that are produced in a cell and that are complementary to a particular mRNA. Antisense RNA can be produced, by inverting the coding region of a gene with respect to its promoter. The antisense RNA can hybridize with its corresponding mRNA, making it double-stranded. The double-stranded mRNA no longer can be recognized by the protein-synthesizing machinery (the ribosomes), and thus expression of this mRNA is suppressed. Also, in many systems double-stranded mRNA is very unstable and is broken down quickly.

Thus, one can inactivate specific genes while not interfering with others. Antisense approaches already are used to protect plants from damage by plant viruses. Reversal of a gene from bean yellow mosaic virus (BYMV), and putting it into tobacco under a reasonably strong promoter, has led to a tobacco variety that is quite resistant to BYMV. A similar approach is used to transfer viral resistance to other plants. This finding is of significance, in that currently no effective, environmentally friendly methods exist to control many plant viruses.

Very related to this approach is the RNAi (RNA interference) approach. This is very useful for both agriculture and medicine, and the first examples of practical applications of this RNAi technology are appearing. As with any new technology, the initial pilot projects are sort of pedantic. However, more exciting application possibilities abound. Obviously, RNAi technology provides an excellent approach for reverse genetics in eukaryotes. With this method one can turn off genes.

Agricultural applications in developing countries

Perhaps indicative of the large potential and relative ease of genetic engineering, developing countries (particularly China) are progressing rapidly in development and application of genetically engineered crops. Some have gone into commercial production well ahead of similar crops in the US. In China, tomatoes that have been engineered for improved virus resistance have been on the market since late 1992.

There are two main reasons for the more rapid commercialization of bioengineered crops in the developing world: (1) less tight governmental approval mechanisms, and (2) hungrier populations. While in developed countries the main value of biotechnological applications may be to reduce production costs, in the developing world a main factor is the production of more food.

Indeed, genetic engineering applications seem to be pretty successful to cut down on pathogen-induced losses. Genetic modification of papaya plants (expression of the ringspot virus coat protein in the plant) protects very well against the very destructive ringspot virus.

ROLE OF GOLGI APPARATUS AND ENDOPLASMIC RETICULUM IN PROTEIN SYNTHESIS

Continued from the previous page that introduces protein synthesis...

mRNA and tRNA

mRNA leaves the nucleus and enters the *cytoplasm* where ribosomes can be found, the site of protein synthesis.

The mRNA strand is met in the ribosome by complimentary tRNA anticodons, which have opposing bases to that of the mRNA strand (the codons).If the mRNA sequence is A-A-U-C-A-U, (codon) then the tRNA sequence is U-U-A-G-U-A (anticodon)

Each *tRNA* molecule consists of 3 bases, deemed an *anticodon* which compliments the opposing bases on the mRNA strand. These in turn have the *amino acid* sequence to successfully code for a particular amino acid.

Each amino acid has a certain sequence of *bases* to make it unique. Therefore, as a summary:

- The initial DNA contained a certain sequence of nucleotides
- The mRNA has a pre-determined sequence (because it is transcribed from the DNA)
- Again, as a consequence the anticodons possess a pre-determined sequence due to the mRNA
- As each amino acid corresponds to a particular anticodon, a unique amino acid sequence is created forming a protein

These amino acids (*peptides*) can combine to form a *polypeptide* chain (proteins), which are used in a variety of structures such as enzymes and hormones.

Ribosomes and Rough Endoplasmic Reticulum (RER)

Ribosomes are the site of protein synthesis, and can occur freely in the cytoplasm though more commonly on the outer surface of rough endoplasmic reticulum. The *endoplasmic reticulum* presents a large surface area on which these ribosomes can be situated, therefore allowing protein synthesis to occur on a large scale.

Rough endoplasmic reticulum is particularly abundant in growing cells which demand a high turnover of materials in its growth. Rough ER is responsible for transporting the newly synthesised proteins to the Golgi apparatus.

The Golgi Apparatus

The *Golgi apparatus* is composed of flattened fluid-filled sacs that controls the flow of molecules in a cell. This is also the case of protein. Carbohydrates are added to the protein to complete its production. This finished product, glycoprotein, is 'pinched off' the Golgi apparatus, and is transported by a vesicle of the cell membrane. When this vesicle reaches the cell membrane, it binds to a receptor on the surface and excretes the protein, where it can then undergo its function.

PHOTOSYNTHESIS - PHOTOLYSIS AND CARBON FIXATION

Photosynthesis is the means that primary producers (mostly plants) can obtain energy via light energy. The energy gained FROM light can be used in various processes mentioned below for the creation of energy that the plant will need to survive and grow.

Photosynthesis is a reduction process, where hydrogen is reduced by a coenzyme. This is in contrast to respiration where glucose is oxidised.

The process is split INTO two DISTINCT areas, *photolysis* (the photochemical stage) and the *Calvin Cycle*. The diagram below gives a summary of the reaction, where light energy is used to initiate the reaction in its presence;

$$CO_2 + H_2O \rightarrow \text{glucose} + \text{oxygen}$$

Photolysis

This part of photosynthesis occurs in the *granum* of *a chloroplast* where light is absorbed by *chlorophyll*; a type of photosynthetic pigment that converts the light to chemical energy. This reacts with water (H_2O) and splits the oxygen and hydrogen molecules apart.

From this dissection of water, the oxygen is released as a by-product while the reduced hydrogen acceptor makes its way to the second stage of photosynthesis, the Calvin cycle.

Overall, since the water is oxidised (hydrogen is removed) and energy is gained in photolysis which is required in the Calvin cycle.

The Calvin Cycle

Also known as the carbon fixation stage, this part of the photosynthetic process occurs in the *stroma* of chloroplasts. The carbon made available FROM breathing in carbon dioxide enters this cycle, which is illustrated below:

Just like the *Kreb's Cycle* in respiration, a substrate is manipulated INTO various carbon compounds to produce energy. In the case of photosynthesis, the following steps occur, which create glucose for respiration FROM the carbon dioxide introduced INTO the cycle;

- Carbon FROM CO_2 enters the cycle combining with Ribulose Biphosphate (RuBP)
- A compound formed is unstable and breaks down FROM its 6 carbon nature to a 3 carbon compound called glycerate phosphate (GP)

- Energy is used to break down GP INTO triose phosphate, while a hydrogen acceptor reduces the compound therefore requiring energy
- Triose Phosphate is the end product of this, a 3 carbon compound which can double up to form glucose, which can be used in respiration.
- The cycle is completed when the leftover GP molecules are met with a carbon acceptor and then turned INTO RuBP, which is to be joined with the carbon dioxide molecules to re-begin the process.

The energy that is used up in the Calvin cycle is the energy that is made available during photolysis. The glucose that is made via GP can be used in respiration or a building block in forming *starch* and *cellulose*, materials that are commonly in demand in plants.

Limiting Factors in Photosynthesis

Some factors affect the rate of photosynthesis in plants, as follows

- *Temperature* plays a role in affecting the rate of photosynthesis. Enzymes involved in the photosynthetic process are directly affected by the temperature of the organism and its environment.
- Light Intensity is also a *limiting factor*, if there is no sunlight, then the photolysis of water cannot occur without the light energy required.
- Carbon Dioxide concentration also plays a factor, due to the supplies of carbon dioxide required in the Calvin cycle stage.

Overall, this is how a plant produces energy which supplies a rich source of glucose for respiration and the building blocks for more complex materials. While animals get their energy FROM food, plants get their energy FROM the sun.

PRINCIPLES OF PLANT GENETICS AND BREEDING

Knowledge of population genetics, quantitative genetics, probability theory and statistics is indispensable for understanding equilibria and shifts with regard to the genotypic composition of a population, its mean value and its variation.

The subject of population genetics is the study of equilibria and shifts of allele and genotype frequencies in populations. These equilibria and shifts are determined by five forces:

- Mode of reproduction of the considered crop

The mode of reproduction is of utmost importance with regard to the breeding of any particular crop and the maintenance of already available varieties. This applies both to the natural mode of reproduction of the crop and to enforced modes of reproduction, like those applied when producing a hybrid variety. In plant breeding theory, crops are therefore classified into the following categories: cross-fertilizing crops, self-fertilizing crops, crops with both cross- and self-fertilization and asexually reproducing crops. It is explained that even within a specific population, traits may differ with regard to their mode of reproduction.

- Selection
- Mutation
- Immigration of plants or pollen, *i.e.* immigration of alleles
- Random variation of allele frequencies

A population is a group of (potentially) interbreeding plants occurring in a certain area, or a group of plants originating from one or more common ancestors. The former situation refers to cross-fertilizing crops (in which case the term Mendelian population is sometimes used), while the latter group concerns, in particular, self-fertilizing crops. In the absence of immigration the population is said to be a closed population. Examples of closed populations are

- A group of plants belonging to a cross-fertilizing crop, grown in an isolated field, *e.g.* maize or rye (both pollinated by wind), or turnips or Brussels sprouts (both pollinated by insects)
- A collection of lines of a self-fertilizing crop, which have a common origin, *e.g.* a single-cross, a three-way cross, a backcross

The subject of quantitative genetics concerns the study of the effects of alleles and genotypes and of their interaction with environmental conditions. Population genetics is usually concerned with the probability distribution of genotypes within a population (genotypic composition), while quantitative genetics considers phenotypic values (and statistical parameters dealing with them, especially mean and variance) for the trait under investigation.

In fact population genetics and quantitative genetics are applications of probability theory in genetics. An important subject is, consequently, the derivation of probability distributions of genotypes and the derivation of expected genotypic values and of variances of genotypic values. Generally, statistical analyses comprise estimation of parameters and hypothesis testing. In quantitative genetics statistics is applied in a number of ways. It begins when considering the experimental design to be used for comparing entries in the breeding programme.

Considered across the entries constituting a population (plants, clones, lines, families) the expression of an observed trait is a random variable. If the expression is represented by a numerical value the variable is generally termed phenotypic value, represented by the symbol p.

Two genetic causes for variation in the expression of a trait are distinguished. Variation controlled by so-called major genes, *i.e.* alleles that exert a readily traceable effect on the expression of the trait, is called qualitative variation. Variation controlled by so-called polygenes, *i.e.* alleles whose individual effects on a trait are small in comparison with the total variation, is called quantitative variation. In Note: it is elaborated that this classification does not perfectly coincide with the distinction between qualitative traits and quantitative traits.

The former paragraph suggests that the term *gene* and *allele* are synonyms. According to Rieger, Michaelis and Green (1991) a gene is a continuous region of DNA, corresponding to one (or more) transcription units and consisting of a particular sequence of nucleotides. Alternative forms of a particular gene are referred to as alleles. In this respect the two terms 'gene' and 'allele' are sometimes interchanged. Thus the term 'gene frequency' is often used instead of the term 'allele frequency'. The term locus refers to the site, alongside a chromosome, of the gene/allele. Since the term 'gene' is often used as a synonym of the term 'locus', we have tried to avoid confusion by preferential use of the terms 'locus' and 'allele' (as a synonym of the word gene) where possible.

In the case of qualitative variation, the phenotypic value p of an entry (plant, line, family) belonging to a genetically heterogeneous population is a discrete random variable. The phenotype is then exclusively (or to a largely traceable degree) a function f of the genotype, which is also a random variable G. Thus $p = f(G)$

It is often desired to deduce the genotype from the phenotype. This is possible with greater or lesser correctness, depending for example on the degree of dominance and sometimes also on the effect of the growing conditions on the phenotype. A knowledge of population genetics suffices for an insight into the dynamics of the genotypic composition of a population with regard to a trait with qualitative variation: application of quantitative genetics is then superfluous.

Note: All traits can show both qualitative and quantitative variation. Culm length in cereals, for instance, is controlled by dwarfing genes with major effects, as well as by polygenes. The commonly used distinction between qualitative traits and quantitative traits is thus, strictly speaking, incorrect. When exclusively considering qualitative variation, *e.g.* with regard to the traits in pea (*Pisum sativum*) studied by Mendel, this book describes the involved trait as a trait showing qualitative variation. On the other hand, with regard to traits where quantitative variation dominates – and which are consequently mainly discussed in terms of this variation – one should realize that they can also show qualitative variation. In this sense the following economically important traits are often considered to be 'quantitative characters':

- Biomass
- Yield with regard to a desired plant product
- Content of a desired chemical compound (oil, starch, sugar, protein, lysine) or an undesired compound
- Resistance, including components of partial resistance, against biotic or abiotic stress factors
- Plant height

In the case of quantitative variation p results from the interaction of a complex genotype, *i.e.* several to many loci are involved, and the specific growing conditions are important. In this book, by complex genotype we mean the sum of the genetic constitutions of all loci affecting the expression of the considered trait. These loci may comprise loci with minor genes (or polygenes), as well as loci with major genes, as well as loci with both. With regard to a trait showing quantitative variation, it is impossible to classify individual plants, belonging to a genetically heterogeneous population, according to their genotypes. This is due to the number of loci involved and the complicating effect on p of (some) variation in the quality of the growing conditions. It is, thus, impossible to determine the number of plants representing a specified complex genotype. (With regard to the expression of qualitative variation this may be possible!). Knowledge of both population genetics and quantitative genetics is therefore required for an insight into the inheritance of a trait with quantitative variation.

The phenotypic value for a quantitative trait is a continuous random variable and so one may write

$$p = f(G,e)$$

Thus the phenotypic value is a function f of both the complex genotype (represented by G) and the quality of the growing conditions (say environment, represented by e). Even in the case of a genetically homogeneous group of plants (a clone, a pure line, a single-cross hybrid) p is a continuous random variable. The genotype is a constant and one should then write

$$p = f(G,e)$$

Regularly in this book, simplifying assumptions will be made when developing quantitative genetic theory. Especially the following assumptions will often be made:

- Absence of linkage of the loci controlling the studied trait(s)
- Absence of epistatic effects of the loci involved in complex genotypes.

These assumptions will now be considered.

Absence of Linkage

The assumption of absence of linkage for the loci controlling the trait of interest, *i.e.* the assumption of independent segregation, may be questionable in specific cases, but as a generalisation it can be justified by the following reasoning. Suppose that each of the n chromosomes in the genome contains M loci affecting the considered trait. This implies presence of n groups of $\binom{M}{2}$_ pairs of loci consisting of loci which are more strongly or more weakly linked. The proportion of pairs consisting of linked loci among all pairs of loci amounts then to

$$\frac{n\binom{M}{2}}{\binom{nM}{2}}=\frac{n.M!}{2!(M-2)!}\times\frac{2!(nM-2)!}{(nM)!}=\frac{M-1}{nM-1}=\frac{1-\frac{1}{M}}{n-\frac{1}{M}}$$

For $M = 1$ this proportion is 0; for $M = 2$ it amounts to 0.077 for rye (*Secale cereale*, with $n = 7$) and to 0.024 for wheat (*Triticum aestivum*, with $n = 21$); for $M = 3$ it amounts to 0.100 for rye and to 0.032 for wheat. For $M \rightarrow \infty$ the proportion is 1 n; *i.e.* 0.142 for rye and 0.048 for wheat.

One may suppose that loci located on the same chromosome, but on different sides of the centromere, behave as unlinked loci. If each of the n chromosomes contains $m\left(=\frac{1}{2}M\right)$ relevant loci on each of the two arms then there are $2n$ groups of $\binom{m}{2}$ _ pairs consisting of linked loci. Thus considered, the proportion of pairs consisting of linked loci amounts to

$$\frac{2n\binom{m}{2}}{\binom{2nm}{2}}=\frac{2n.m!}{2!(m-2)!}\times\frac{2!(2nm-2)!}{(2nm)!}=\frac{1-\frac{1}{m}}{2n-\frac{1}{m}}$$

For $m = 1$ this proportion is 0; for $m = 2$ it amounts to 0.037 for rye and to 0.012 for wheat; for $m = 3$ it amounts to 0.049 for rye and to 0.016 for wheat. For $m \rightarrow \infty$ the proportion is $\frac{1}{2n}$.*e.* 0.071 for rye and 0.024 for wheat.

For the case of an even distribution across all chromosomes of the polygenic loci affecting the considered trait it is concluded that the proportion of pairs of linked loci tends to be low. (In an autotetraploid crop the chromosome number amounts to $2n = 4x$. The reader might like to consider what this implies for the above expressions.)

ABSENCE OF EPISTASIS

Absence of epistasis is another assumption that will be made regularly in this book. It implies additivity of the effects of the single-locus genotypes for the loci affecting the level of expression for the considered trait. The genotypic value of some complex genotype consists then of the sum of the genotypic value of the complex genotype with regard to all non-segregating loci, here represented by m, as well as the sum of the contributions due to the genotypes for each of the K segregating polygenic loci $B1$-$b1$,..., BK-bK. Thus

$$GB_1-b_1....,B_k-b_k=m+G'B_1-b_1+...+G'B_k-b_k$$

where $G_$ is defined as the contribution to the genotypic value, relative to the population mean genotypic value, due to the genotype for the considered locus. The assumption implies the absence of inter-locus interaction, *i.e.* the absence of epistasis (in other words: absence of non-allelic interaction). It says that the effect of some genotype for some locus $Bi - bi$ in comparison to another genotype for this same locus does not depend at all on the complex genotype determined by all other relevant loci.

In this book, in order to clarify or substantiate the main text, theoretical examples and results of actual experiments are presented. Notes provide short additional information and appendices longer, more complex supplementary information or mathematical derivations.

3

Effects of Mating System and Genetic Variability

POPULATION GENETIC EFFECT OF SELECTION WITH REGARD TO SEX EXPRESSION

The types of sex expression distinguished for our purposes are

- Hermaphroditism, in contrast to
- Sex differentiation (sexual dimorphism)

Hermaphroditism is the most common form of sex expression among plant species. It means that the reproductive organs of both sexes are present in the same flower, *i.e.* a bisexual flower (this situation is indicated by the symbol), or in different flowers occurring on the same plant. In the latter case a flower contains either male or female organs; this situation is called monoecy, indicated by the symbol Band@and. Monoecy occurs in crops such as

Maize *Zea mays* L.
Castorbean *Ricinus communis* L.
Cucumber *Cucumis sativus* L.
Plane trees *Platanus occidentalis* L.
Alder *Alnus glutinosa* Gartn.
Hazelnut *Corylus avellana* L.

The types of sex differentiation to be distinguished are

- Dioecy
- Gynodioecy

Dioecy means that plants either exclusively produce female flowers (these are female plants), or exclusively male flowers (these are the male plants). Well-known dioecious crops are

Spinach *Spinacia oleracea* L.
Asparagus *Asparagus officinalis* L.
Hemp *Cannabis sativa* L.
Hops *Humulus lupulus* L.
Poplar *Populus nigra* L.

Date *Phoenix dactylifera* L.
Kiwi *Actinidia deliciosa*
Papaya *Carica papaya* L.

Gynodioecy means that female plants as well as hermaphroditic plants occur. Thus a gynodioecious maize population consists of male sterile plants, *i.e.* female plants, as well as 'normal' plants.

It has been demonstrated that sex expression, both in plants and animals, is due to rather diverse mechanisms, ranging from a more or less clear-cut *XY* - *XX*-mechanism to sex expression determined by environmental conditions.

Example. In cucumber four types of sex expression may occur: monoecy, gynoecy, and andromonoecy (plants have male and hermaphroditic flowers) and hermaphroditism. Modern cucumber cultivars produce exclusively female flowers: their fruits develop parthenocarpic.

The sex expression is affected by treatment with gibberellic acid or silvernitrate. These substances promote the development of male flowers. This allows the selfing required for maintenance of pure lines used in hybrid varieties. The population genetic effect of selection with regard to sex expression is thus necessarily derived on the basis of simplifying assumptions about the genetic control of sex expression. In this chapter implications of specific assumptions about the genetic control of dioecy or gynodioecy are elaborated.

ASSUMED GENETIC CONTROL OF DIOECY

A 'homozygous' genotype is assumed to give rise to a female plant, *viz.* *XX* in the case of sex chromosomes or *mm* in the case of a locus *M-m* controlling sex expression. A 'heterozygous' genotype (*XY* or *Mm*) is assumed to give rise to a male plant:

Genotype

	mm (or: XX)	Mm (or: XY)
sex	♀	♂
f	$\frac{1}{2}$	$\frac{1}{2}$

The genotypic composition (1/ 2, 1/ 2, 0) results from the harvesting of female plants which have been pollinated by male plants. This genotypic composition will apply whatever the initial frequencies of male and female plants.

ASSUMED GENETIC CONTROL OF GYNODIOECY

Gynodioecy occurs in the situation of cytoplasmic male sterility or in the situation of genic male sterility. The idiotypic basis for cytoplasmic male sterility is assumed to be

Idiotype	(S)rr	(.)Rr	(.)RR
Sex	♀	☿ or ⚥	☿ or

The symbol (S) designates presence of male-sterility-inducing cytoplasm, the symbol (·) presence of any cytoplasm. The latter symbol represents thus both (S) and (N), *i.e.* the presence of normal cytoplasm. Locus *R-r* is the male fertility restoring locus.

The genetic basis for genic male sterility is assumed to be

Genotype	mm	Mm	MM
Sex	♀	⚥ or ♂	⚥ or ♂

In the case of gynodioecy there is selection against the male-sterility-inducing allele (this is allele *m*; or – in the presence of (S) cytoplasm – allele *r*). Male sterile plants are unable to transmit this allele to the next generation via pollen.

THE FREQUENCY OF MALE STERILE PLANTS

Allogamous Crops

In cross-fertilizing crops male sterile plants may have a normal (complete) seed set. The selection against the male-sterility-inducing allele, say *m*, is then due to the incapability of plants with genotype *mm* to transmit allele *m* via the pollen to the next generation. Only plants with genotype*MM* or*Mm*produce pollen.

Eggs are produced by all plants, whatever the genotype. The frequency of male sterile plants in this situation is considered. Elimination of male sterility may be a breeding objective because of a low seed-set on the male sterile.

Male sterile plants, which may be conspicuous because of their low seed-set, are then not harvested. This implies that plants with genotype *mm* not only fail to produce pollen, but – effectively – then also fail to produce eggs.

Only male fertile plants are harvested. In successive generations the genotypic composition with regard to locus *M-m* coincides then with the genotypic composition with regard to locus *A-a* in the case of continued mass selection, before pollen distribution, against plants with genotype *aa*. The decrease in the frequency of gene *m* proceeds, therefore, as in Example.

Autogamous Crops

Incomplete seed-set is certainly to be expected for male sterile plants belonging to a self-fertilizing crop. Attention is given to natural selection against male sterility in an autogamous crop. In the case of recurrent selection in a self-fertilizing crop, only male sterile plants are harvested. This guarantees that the harvested seeds resulted from intercrossing. Then, effectively, plants with genotype *MM* or *Mm* produce the pollen and plants with genotype *mm* the eggs. This situation coincides effectively with dioecy. It leads immediately to the equilibrium frequencies (1/ 2, 1/ 2, 0), whatever the seed-set of male sterile plants may be.

COMPLETE SEED-SET OF THE MALE STERILE PLANTS

The situation of complete seed-set of male sterile plants of a cross-fertilizing crop resembles the case of mass selection, after pollen distribution, against plants with genotype *aa*: such plants are not harvested and, consequently, do not transmit allele *a* via eggs; pollen, however, is produced by all plants, whatever the genotype. In successive generations the genotypic composition with regard to locus *M-m* is, consequently, equal to the genotypic composition with regard to locus *A-a* in the case of mass selection, after pollen distribution, against plants with genotype *aa*. This is illustrated in Example.

Consider now a gynodioecious population of a cross-fertilizing crop, *e.g.* maize: female plants have idiotype (S)*rr* and hermaphroditic plants idiotype (N)*rr*. The relative frequencies of female plants and hermaphroditic plants will then not change if these two categories of plants have equal seed-set. The problem described in Note pertains to this situation.

INCOMPLETE SEED-SET OF THE MALE STERILE PLANTS

In the case of cytoplasmic male sterility in a self-fertilizing crop the incomplete seed-set of the male sterile plants, due to insufficient pollination, implies reduction of the frequency of plants with the (S) cytoplasm. With cleistogamy, *i.e.* the flowers remain closed at pollination time, there is no seed-set at all.

Plants with the (S) cytoplasm do then not produce any offspring. The (S) cytoplasm will then not be transmitted to the next generation. It is immediately lost. In the remainder of this section attention is given to genic male sterility in a self-fertilizing crop. It is assumed that all seeds produced by hermaphroditic plants, *i.e.* by plants with genotype *Mm* or *MM*, are due to self-fertilization.

For these plants the value for k, *i.e.* the portion of the eggs that develop into a zygote after cross-fertilization is zero. The seeds produced by male sterile plants, *i.e.* plants with genotype *mm*, are due to cross-fertilization. It is rather common that male sterile plants produce flowers that are more widely opened than flowers produced by male fertile plants, but nevertheless they tend to produce less seeds than male fertile plants.

The relative seedset or – in more general population genetic terms – the relative fitness of plants with genotype *mm* is represented by the factor $w0$. (The relative fitness is also designated by $1 - s0$, or briefly by $1 - s$, where s represents the so-called selection coefficient for plants with genotype*mm*). Example Even for a crop like spring barley, k appears to be positive. Jain and Allard (1960) observed $k = 0.02$ for hermaphroditic barley plants. The seed-set of male sterile barley plants is rather variable. For the conditions in Davis, California, Jain and Suneson (1964) reported a maximum seedset of 0.40; *i.e.* s $e-.6$. For Wageningen, The Netherlands, Baltjes (1975) reported a maximum seed-set of 0.20; *i.e.* s $e- 0.8$.

Different parental genotypes produce different numbers of offspring. The effective (relative) frequencies (*fe*) of parental genotypes are calculated from their actual frequencies in the following way:

Genotype

	mm	*Mm*	*MM*
f	$f_{0,t}$	$f_{1,t}$	$f_{2,t}$
w	$1-8$	1	1
f_e	$\dfrac{(1-s)f_{0,t}}{1-sf_{0,t}}$	$\dfrac{f_{1,t}}{1-sf_{0,t}}$	$\dfrac{f_{2,t}}{1-sf_{0,t}}$

Plants with genotype *Mm* or *MM* are assumed to produce offspring by spontaneous self-fertilization:

- The genotypic composition of the offspring of plants with genotype *Mm* is (1/ 4, 1/ 2, 1/ 4).
- The genotypic composition of the offspring of plants with genotype *MM* is (0, 0, 1).

Plants with genotype *mm* produce offspring by cross-fertilization. The haplotypic composition of the pollen produced by generation t is

Haplotype

	m	*M*
f	$g_{0,t+1}$	$g_{1,t+1}$

Where

$$g_{0,t+1} = \frac{\frac{1}{2}f_{1,t}}{1-f_{0,t}} \textit{ and } g_{1,t+1} = \frac{\frac{1}{2}f_{1,t}+f_{2,t}}{1-f_{0,t}}$$

The genotypic composition of the offspring of plants with genotype *mm* is ($g0,t+1$, $g1,t+1$, 0). Altogether the genotypic composition of generation $t+1$, in terms of the genotype frequencies in generation t is

Genotype

	mm	*Mm*	*MM*
f	$\dfrac{\dfrac{\frac{1}{2}f_{1,t}(1-s)f_{0,t}}{1-f_{0,t}}}{1-sf_{0,t}}$	$+\dfrac{1}{4}f_{1,t}\quad \dfrac{\left(\frac{1}{2}f_{1,t}+f_{2,t}\right)\dfrac{(1-s)f_{0,t}}{1-f_{0,t}}}{1-sf_{0,t}}+$	$\dfrac{\frac{1}{2}f_{1,t}+f_{2,t}}{1-sf_{0,t}}$

The frequency of plants with genotype *Mm* decreases due to self-fertilization but, on the other hand, it increases due to cross-fertilization of plants with genotype *mm*. The frequency of plants with genotype *MM* can only increase. The eventual genotypic composition is thus (0, 0, 1). This limit is approached more quickly when the seed-set of plants with genotype *mm* is lower, *i.e.* s is larger. Example illustrates the reduction of *f*0 for a few values for s. Example Table presents *f*0, *i.e.* the frequency plants with genotype *mm*. It does so for several values of s and for successive generations, starting with an initial population with the genotypic composition of an F2, *i.e.* (1/ 4, 1/ 2, 1/ 4). The column headed by '$s = 0$' represents complete seed-set of male sterile plants. The column headed by '$s = 1$', representing complete sterility, illustrates how *f*0 is reduced by mass selection in a self-fertilizing crop against plants with genotype *mm*. The column headed 'Observed frequency' presents actual data obtained from barley, Composite Cross XXI. The frequencies presented in this column and in the column headed '$s = 0.8$' are depicted in Figure. It appears that *f*0 decreased in later generations less than calculated for $s = 0.8$: from population F8 onward the actual values for *f*0 were somewhat higher than the calculated values. Some tentative explanations for this are given at the end of the present section.

Suneson (1956) advocated the so-called *evolutionary plant breeding method*. It is based on the thought that natural selection in a genetically heterogeneous population favours, for certain traits, the same phenotypes as preferred by the breeder. The improvement of the population will be slow, but in the long run sufficient for obtaining attractive plant material. Example provides some results.

Table: The (expected) frequency of male sterile plants (with genotype mm) in successive generations. The genotypic composition of the initial population is The relative fitness of the male sterile plants is 1-s. The column headed by 'Observed frequency' present actual data obtained from barley

Population	Frequency of male sterile plants expected for $s = 0$	$s = 0.6$	$s = 0.8$	$s = 1$	Observed frequency
F_2	0.250	0.250	0.250	0.250	
F_3	0.208	0.186	0.177	0.167	
F_4	0.159	0.124	0.122	0.100	0.060
F_5	0.125	0.082	0.069	0.056	
F_6	0.098	0.054	0.042	0.029	0.037
F_7	0.078	0.035	0.025	0.015	0.023
F_8	0.062	0.023	0.015	0.008	0.020
F_9		0.016	0.009		0.010
F_{10}		0.010	0.005		0.013
F_{11}		0.003			
F_{12}		0.002			0.010
F_{13}		0.001			0.006

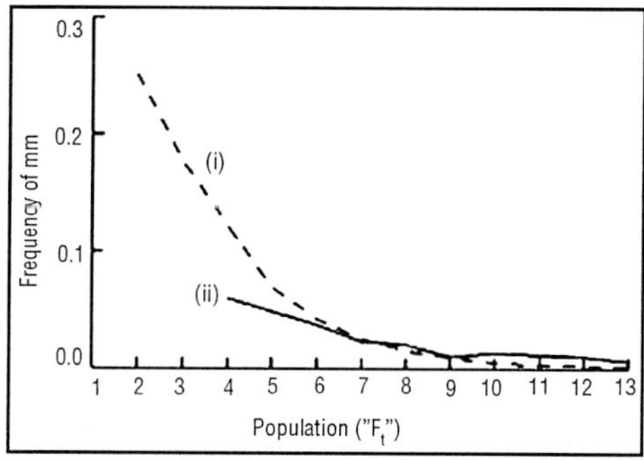

Fig. The Frequency of Male Sterile Plants, with Genotype mm, in Successive Generations.

The genotypic composition of the original population was (1/4,1/2,1/4). (i) Data calculated for a relative fitness of the male sterile plants equal to 1-s=0.2, and (ii) Observed data in barley

Example. To test the *'evolutionary plant breeding method'* hypothesis, Suneson developed broad base populations by open pollination of male sterile lines. He developed Composite Cross XXI by growing 6200 spring barley varieties next to male sterile barley plants. The seed harvested from the male sterile plants was used as the source population.

This population was grown for many years/generations. Baltjes (1975) studied, within the same growing season, many generations, as derived in Wageningen, The Netherlands. A significant improvement in resistance to powdery mildew appeared. As for yield, however, no clear effect was observed: relative to the check variety Zephyr, the F4 population yielded 75.7per cent and the F13 population 83.7per cent.

Baltjes (1975) observed that $f0$ decreased in later generations less than calculated for $s = 0.8$: from F8 onward the actual frequency of plants with genotype *mm* was somewhat higher than the calculated frequency. Two tentative explanations are presented:

- The relative fitness of male sterile plants may increase in the course of the generations. Thus seed-set improves. This could be due to more intense pollination because of the increase in the frequency of male fertile plants. Indeed, Jain and Suneson (1964) reported a seed-set of 40per cent in generation F18 and a seed-set of 60per cent in generation F21. They, therefore, assumed a higher relative fitness of male sterile plants at a lower frequency of such plants: $1 - s$ was taken to be $0.6 - f0$.
- Male sterile plants (genotype *mm*) produce offspring heterozygous for many loci. Due to this highly heterozygous background-genotype these offspring (genotype *mm* or *Mm*), may tend to be more vigorous than the more homozygous plants (genotype *mm*, *Mm* or *MM*) obtained after selfing. Constancy of q, the frequency of gene m, may occur if its potential decrease, because of reduced fertility of *mm* plants, is offset by its potential increase, due to greater vitality of *mm* plants belonging to the heterozygous offspring of plants with genotype *mm* (Jain and Suneson, 1964).

AUTOTETRAPLOID CHROMOSOME BEHAVIOUR AND PANMIXIS

The implications of panmixis in an autotetraploid crop will only be considered for a single locus with two alleles. This is to keep the mathematical derivations simple. It will be shown that the equilibrium frequencies of the genotypes are not obtained after a single panmictic reproduction. At equilibrium the frequencies of the genotypes and the haplotypes are equal to the products of the frequencies of the alleles involved. Among cross-fertilizing autotetraploid crops the more important representatives are alfalfa (*Medicago sativa* L.; $2n = 4x = 32$) and cocksfoot (*Dactylis glomerata* L.; $2n = 4x = 28$). Additionally, highbush blueberry (*Vaccinium corymbosum* L.; $2n = 4x = 48$) might be mentioned. Leek (*Allium porrum* L.; $2n = 4x = 32$) is an autotetraploid crop with a tendency to a diploid behaviour of the chromosomes. Among ornamentals several autotetraploid species occur, *e.g.* *Freesia hybrida, Cyclamen persicum* Mill. ($2n = 4x = 48$) and *Begonia semperflorens*. Also, artificial autotetraploid crops have been made, *e.g.* rye (*Secale cereale* L.; $2n = 4x = 28$) and perennial rye grass (*Lolium perenne* L.; $2n = 4x = 28$). In 1977 about 500,000 ha of autotetraploid rye were grown in the former Soviet Union. Sweet potato, *i.e.* *Ipomoea batatas* var. *littoralis* ($2n = 4x = 60$) or *I. batatas* var. *batatas* ($2n = 6x = 90$), may be considered as a cross-fertilizing crop (due to self-incompatibility), but it is mainly vegetatively propagated.

Under certain conditions double reduction may occur in autotetraploid crops, in which case (parts of) sister chromatids end up in the same gamete.

The resulting haplotype is homozygous for the loci involved. The process of double reduction causes the frequency of homozygous genotypes and haplotypes to be somewhat higher than in absence of double reduction. Blakeslee, Belling and Farnham (1923) discovered the phenomenon in autotetraploid jimson weed (*Datura stramonium* L.; $2n = 4x = 48$): a triplex plant (with genotype *AAAa*) produced some nulliplex offspring after crossing with a nulliplex (genotype *aaaa*). This is only possible if the triplex plant produces *aa* gametes. The process of double reduction is an interesting phenomenon, but in a quantitative sense it is of no importance. For this reason we assume that double reduction does not occur. The autotetraploid genotypes to be distinguished for locus *A-a* are *aaaa* (nulliplex), *Aaaa* (simplex), *AAaa* (duplex), *AAAa* (triplex) and *AAAA* (quadruplex).

In each cell these genotypes contain $J A$ alleles and $4 - J a$ alleles. At meiosis two of these four alleles are sampled to produce a gamete. The haplotypes that can be produced by an autotetraploid plant containing $J A$ alleles can be described by j, the number of A alleles that they contain, where $j = 0$, 1 or 2. The conditional probability distribution for j, given that the parental genotype contains $J A$ alleles, is a hypergeometric probability distribution:

$$P(\underline{j}=j/J)=\frac{\binom{J}{j}\binom{4-J}{2-j}}{\binom{4}{2}}=\frac{1}{6}\binom{J}{j}\binom{4-J}{2-j}$$

The Probability that a triplex plant (*i.e.*j =3) Produces a gemete with haplotype Aa (*i.e.* j = 1) is therefore

$$P(\underline{j}=1/J=3)=\frac{1}{6}\binom{3}{1}\binom{1}{1}=\frac{1}{2}$$

Table Presents, for each autotetraploid genotype, the haplotypic composition, *i.e.* the probability distribution for the haplotypes produced. The genotypic composition of a tetraploid population is described like that of a diploid population. Thus in the case of autotetraploid species the row

Table. The haplotypic composition of the gametes produced by each of the five autotetraploid genotypes that can be distinquished for locus A-a

	Haplotype		
Genotype	*aa*	*Aa*	*AA*
aaaa	1	0	0
Aaaa	$\frac{1}{2}$	$\frac{1}{2}$	0
AAaa	$\frac{1}{6}$	$\frac{4}{6}$	$\frac{1}{6}$
AAAa	0	$\frac{1}{2}$	$\frac{1}{2}$
AAAA	0	0	1

Vector (*f*0, *f*1, *f*2, *f*3, *f*4) is used. The equilibrium frequencies of the genotypes are attained as soon as the haplotype frequencies are stable. Therefore the haplotypic composition of successive generations with panmictic reproduction will be monitored.

Some initial population G0 produces gametes with haplotypic composition:

	Haplotype		
	aa	*Aa*	*AA*
f	*g*0,1	*g*1,1	*g*2,1

The frequency of a is

And that of A is

$$p=\frac{1}{2}g1,1+g2,1$$

Panmictic reproduction of G_0 yields population G_1 with the following genotypic composition:

Genotype

	aaaa	*Aaaa*	*AAaa*	*AAAa*	*AAAA*
f	$g0,1^2$	$2g0,1g1,1$	$g1,1^2+2g0,1g2,1$	$2g1,1g2,1$	$g2,1$

The haplotypic composition of the gametes produced by G_1 is:

Haplotype

	aa	*Aa*	*AA*
f	$g0,2$	$g1,2$	$g2,2$

According to table the following applies:

$$g1,2=\frac{1}{2}(2g0,1g1,1)+\frac{2}{3}(g1,1^2+2g0,1g2,1)+\frac{1}{2}(2g1,1g2,1)$$

$$=\frac{2}{3}(\frac{3}{2}g0,1g1,1+\frac{3}{2}g1,1g2,1+g1,1^2+2g0,1g2,1)$$

$$=\frac{2}{3}[2(g0,1+\frac{1}{2}g1,1)(\frac{1}{2}g1,1+g2,1)+\frac{1}{2}g1,1(g0,1+g1,1+g2,1)]$$

$$=\frac{2}{3}(2pq+\frac{1}{2}g1,1)$$

Generally

$$g1,t+1=\frac{2}{3}(2pq+\frac{1}{2}g1,t)$$

The frequencies of the genotypes have attained their equilibrium (*e*) values as soon as the frequencies of the haplotypes are constant. The latter implies:

$$g1,e=\frac{2}{3}(2pq+\frac{1}{2}g1,e),$$

i.e.

$$g1,e=2pq$$

The haplotype frequencies are then

$$g0,e=q-\frac{1}{2}g1,e=q-pq=q^2$$

$$g1,e=2pq$$

$$g2,e=p-\frac{1}{2}g1,e=p-pq=p^2$$

The genotypic composition in equilibrium is consequently

Genotype

	aaaa	*Aaaa*	*AAaa*	*AAAa*	*AAAA*
f	q^4	$4pq^3$	$6p^2q^2$	$4p^3q$	p^4

This composition is also given by the probability distribution for *J*, the number of A alleles in the autotetraploid genotype:

$$P(\underline{j}=J)=\binom{4}{J}p^J q^{4-J}$$

The deviation from the equilibrium is measured by the quantity *dt*, which measures the excess or deficit of the frequency of gametes with the *Aa* haplotype with regard to their equilibrium frequency. Thus *dt* is defined as follows:

$$d_t := g1,t - g1,e$$

The rate of decrease of *dt* indicates how fast the equilibrium is approached. Equations Yield

$$d_{t+1} = g1,t+1 - g1,e = \frac{2}{3}(2pq + \frac{1}{2}g1,t) - 2pq = \frac{1}{3}(g1,t - g1,e) = \frac{1}{3}dt$$

One round of panmictic reproduction produces a population in which the deviation amounts only to 1/ 3 of the deviation in the preceding population.

The equilibrium is approached in an asymptotic way. Example gives an illustration.

Example.The approach of the equilibrium is considered for an initial population G0 with genotypic composition (0.04; 0; 0.72; 0; 0.24). The haplotype frequencies are:

$$g0,1 = 0.04 + 0.12 = 0.16$$
$$g1,1 = 0.48$$
$$g2,1 = 0.12 + 0.24 = 0.36$$

Thus q = 0.4 and p = 0.6 This implies that:

$$g0,1 = q^2 = g0,e$$
$$g1,1 = 2pq = g1,e$$
$$g2,1 = p^2 = g2,e$$

For a more advanced treatment of the population genetic theory of crossfertilizing crops with an autotetraploid behaviour of the chromosomes the reader is referred to Seyffert (1960). Finally, it is emphasized once again that in this section it was assumed that the population contains only two different alleles for the segregating locus.

In fact more alleles may occur in such a way that plants with three or four different alleles per locus are present, *viz.* plants with genotype *AiAiAjAk* or *AiAjAkAl*, respectively. Quiros (1982) reported such genotypes for isozyme loci in alfalfa. Some claims have been made that plants with a heterozygous genotype containing three or four different alleles for the considered locus, are more vigorous than plants with a heterozygous genotype containing one or two alleles.

MORE THAN TWO LOCI, EACH WITH TWO OR MORE ALLELES

Linkage Involving Three Loci

Three loci *A-a*, *B-b* and *C-c* are considered. These loci occur in this order along a chromosome. The segments *AB*,*BC* and *AC* are distinguished. Effective recombination of alleles belonging to loci *A-a* and *B-b* requires that the number of crossover events in segment *AB* is an odd number. The probability of recombination is called recombination value, designated by the symbol *rc,* or by the symbol *rAB* or simply by *r* (depending on the context). With an even number of times of crossing-over in segment *AB* there is no (effective) recombination. The probability of this event is $1 - rAB$. There is (effective) recombination of alleles belonging to loci *A-a* and *C-c* if there is either (effective) crossing-over in segment *AB*, but not in segment BC; or if there is (effective) crossing-over in segment *BC*, but not in segment *AB*.

If the occurrence of recombination in one chromosome segment has no effect on the recombination value for the adjacent segment the following relation applies:

$$rAC = rAB(1 - rBC(1 - rAB) = rAB + rBC - 2rABrBC$$

This situation is likely for loci that are not too closely linked. The situation where recombination in one segment depresses the probability of recombination in an adjacent segment is called chiasma interference. A more general expression for *rAC* is thus:

$$rAC = rAB + rBC - 2(1 - \delta)rABrBC$$

where δ is the interference parameter, ranging from 0 (no interference) through 1 (complete interference). It shows that *rAC* is higher at higher values for δ. Recombination values are additive if

$$2(1 - \delta)rABrBC = 0$$

i.e. if $\delta = 1$ and/or $rABrBC = 0$. In other cases they are not additive. These conditions imply that recombination values are mostly not additive. They are, consequently, inappropriate to measure distances between loci.

The hypothesis of independence of crossing-over in segments *AB* and *BC*, *i.e.* the hypothesis of absence of chiasma interference, can be tested by means of a goodness-of-fit test. Among *N* plants, the expected number of plants with a genotype which is due to double crossing-over amounts, according to this hypothesis, to *rABrBCN*.

It is compared to the observed number. The ratio

$$\frac{\textit{Observed Number}}{\textit{Expected Number}}$$

is called coefficient of coincidence. When there is independency it is equal to 1. Its complement, *i.e.*

$$1-\frac{Observed\ Number}{Expected\ Number}$$

Estimates δ. Its value is positive if the observed number of plants with the recombinant genotype is smaller than the number expected at independency: the presence of a chiasma in the one segment hinders the formation of a chiasma in the other segment.

The actual distance between loci, say the map distance m, measures the total number of cross-over events (both odd and even numbers) between the loci. This distance is an additive measure. It can only approximately be determined from recombination values. Haldane (1919) developed an approximation for the situation in the absence of interference ($\delta = 0$). His mapping function is

$$m=-\frac{In(1-2r_c)}{2},$$

Where m represents the expected number of cross-over events in the considered segment (Kearsey and Pooni, 1996; pp. 127–130). As the map distance is mostly expressed in centiMorgans (cM), this function is

$$m=-50\ln(1-2r_c)$$

Often written as An approximation which takes interference into account is called Kosambi's mapping function.

Frequencies of Complex Genotypes after Continued Panmixis

It can be shown that continued panmixis eventually leads to an equilibrium of the frequencies of complex genotypes for three or more loci, each with two or more alleles. The equilibrium is characterized by haplotype frequencies equal to the products of the frequencies of the alleles involved. Linkage equilibrium for one or more pairs of loci does not imply equilibrium of the frequencies of complex genotypes for three or more loci. Equilibrium of the frequencies for complex genotypes implies, however, linkage equilibrium for all pairs of loci.

TWO LOCI, EACH WITH TWO ALLELES

Iit is shown that a single round of panmictic reproduction produces immediately the Hardy–Weinberg genotypic composition with regard to a single locus. It is immediately attained because the random fusion of pairs of gametes implies random fusion of separate alleles, whose frequencies are constant from one generation to the next. For complex genotypes, *i.e.* genotypes with regard to two or more loci (linked or not), however, the so-called linkage equilibrium is only attained after *continued panmixis*. Presence of the Hardy–Weinberg genotypic composition for separate loci does not imply presence of linkage equilibrium! (Example illustrates an important exception to this rule.)

In panmictic reproduction the frequencies of complex genotypes follow from the frequencies of the complex haplotypes. Linkage equilibrium is thus attained if the haplotype frequencies are constant from one generation to the next. For this reason 'linkage equilibrium' is also indicated as gametic phase equilibrium. In this section it is derived how the haplotypic frequencies approach their equilibrium values in the case of continued panmixis. This implies that the tighter the linkage the more generations are required. However, even for unlinked loci a number of rounds of panmictic reproduction are required to attain linkage equilibrium.

The genotypic composition in the equilibrium does not depend at all on the strength of the linkage of the loci involved. The designation 'linkage equilibrium' is thus not very appropriate.

To derive how the haplotype frequencies approach their equilibrium. We consider loci A-a and B-b, with frequencies p and q for alleles A and a and frequencies r and s for alleles B and b. The recombination value is represented by rc. This parameter represents the probability that a gamete has a recombinant haplotype. Independent segregation of the two loci occurs at $rc = 1/2$, absolute linkage at $rc = 0$. Example illustrates the estimation of rc in the case of a testcross with a line with a homozygous recessive (complex) genotype.

The haplotype frequencies are determined at the meiosis. The haplotypic composition of the gametes produced by generation Gt–1 is described by

Haplotype			
ab	*aB*	*Ab*	*AB*
$g_{00,t}$	$g_{01,t}$	$g_{10,t}$	$g_{11,t}$

The last subscript (t) in the symbol for the haplotype frequencies indicates the rank of the generation to be formed in a series of generations generated by panmictic reproduction (t = 1, 2,...).

Example. The spinach variety Wintra is susceptible to the fungus *Peronospora spinaciae* race 2 and tolerant to Cucumber virus 1. It was crossed with spinach variety Nores, which is resistant to *P. spinaciae* race 2 but sensitive to Cucumber virus 1.

The loci controlling the host-pathogen relations are $A - a$ and $B - b$. The genotype of Wintra is *aaBB* and the genotype of Nores *AAbb*. The offspring, with genotype *AaBb*, were crossed with the spinach variety Eerste Oogst (genotype *aabb*), which is susceptible to *P. spinaciae* race 2 and sensitive to Cucumber virus 1. On the basis of the reaction to both pathogens a genotype was assigned to each of the 499 plants resulting from this testcros:

Genotype				
aabb	aaBb	Aabb	AaBb	Total
61	190	194	54	499

124.75 124.75124.75124.75499

The expected frequencies are calculated on the basis of the null hypothesis stating that the two involved loci are unlinked. The expected 1 2: 1 2 segregation ratio was confirmed by a goodness of fit test for each separate locus. The specified null hypothesis is, of course, rejected. The two loci are clearly linked. The value estimated for *rc* is

$$\frac{61+54}{499}=0.23$$

Note: In this book the last subscript in the symbols for the genotype and haplotype frequencies indicate the generation number. If it is *t* it refers to population G*t*, *i.e.* the population obtained by panmictic reproduction of *t* successive generations.

Population G1, resulting from panmictic reproduction in a single-cross hybrid, has the same genotypic composition as the F2 population resulting from selfing plants of the single-cross hybrid. To standardize the numbering of generations of cross-fertilizing crops and those of self-fertilizing crops, the population resulting from the first reproduction by means of selfing might be indicated by S1 (rather than by the more common indication F2).

To avoid confusion this will only be done when appropriate. The last subscript in the symbols for the haplotype frequencies of the gametes giving rise to S1 are taken to be 1. The same applies to the frequencies of the genotypes in S1. This system for labelling generations of gametophytes and sporophytes was also adopted by Stam.

Population G0 is thus some initial population, obtained after a bulk cross or simply by mixing. It produces gametes with the haplotypic composition $(g00,1; g01,1; g10,1; g11,1)$.

In the absence of selection, allele frequencies do not change. This implies

$$g_{10,1}+g_{11,1}=g_{10,2}+g_{11,2}=\dots=p$$

for allele *A*, and similar equations for the frequencies of alleles *a*,*B* and *b*.

It was already noted that the haplotype frequencies in successive generations will be considered. In the appendix of this section it is shown that the following recurrent relations apply:

$$g_{100,t}+1=g_{100,t}-r_c d_t$$

$$g_{01,t}+1=g_{01,t}+r_c d_t$$

$$g_{10,t}+1=g_{10,t}+r_c d_t$$

$$g_{11,t}+1=g_{11,t}-r_c d_t$$

Where the definition of d_t follows from:

$$2d_t := f_{11c,t}-f_{11R,t}$$

where ':=' means: 'is defined as', and t = 1, 2, 3,...

In Note: it is shown that Equations also apply to selffertilizing crops. The recurrent equations show that the haplotype frequencies do not change from one generation to the next if $rc = 0$ or if $dt = 0$. Such constancy of the haplotypic composition implies constancy of the genotypic composition. It implies presence of linkage equilibrium. Linkage equilibrium is thus immediately established by a single round of panmictic reproduction for loci with $rc = 0$. This situation coincides with the case of a single locus with four alleles.

The symbol $f11C$ indicates the frequency of *AB*/*ab*-plants, *i.e.* doubly heterozygous plants in coupling phase (C-phase); the symbol $f11R$ represents the frequency of *Ab*/*aB*-plants, *i.e.* doubly heterozygous plants in repulsion phase (R-phase).

In the case of panmixis the following equations apply:

$$f11c,t = 2(g11,tg00,t)$$

$$f11r,t = 2(g10,tg01,t)$$

In that case we get

$d_t = (g11,tg00,t) - (g10,tg01,t)$

This parameter is called coefficient of linkage disequilibrium. It appears in the following derivation:

$g_{11},t = g_{11},t(g_{10},t + g_{01},t + g_{11},\ t + g_{00},t) = (g_{10},tg_{01},t + g_{10},tg_{11},t$
$+g_{11},tg_{01},t + g_{11},t) + (g_{11},tg_{00},t - g_{10},tg_{01},t)$
$= (g_{10},t + g_{11},t)(g_{01},t + g_{11},t) + d_t = p_r + d_t$

Equation may thus be rewritten as

$pr + d_{t+1} = (pr + d_t) - r_c d_t$

Which implies not only

$d_{t+1} = (1 - r_c)d_t$

equation may thus be rewritten as

but of course also

$d_t = (1 - r_c)^{t-1} d_1$

for $t = 2,3,\ldots$.

The derivation above (and similar derivations for the other haplotype frequencies)implies

$d_t = g11,t - pr = -(g10,t - ps) = -(g01,t - qr) = g00,t - qs$

Because $1/2 \leq (1{-}rc) \leq 1$, continued panmixis implies continued decrease of dt. The decrease is faster for smaller values of $1 - rc$, *i.e.* for higher values of rc. Independent segregation, *i.e.* $rc = 1/2$, yields the fastest reduction, *viz.* halving of dt by each panmictic reproduction. The value of dt eventually attained, *i.e.* $dt = 0$, implies that linkage equilibrium is attained, *i.e.* constancy of the haplotype frequencies. The haplotype frequencies have then a special value, *viz.*

$$g00 = qs$$
$$g01 = qr$$
$$g10 = ps$$
$$g11 = pr$$

The equilibrium frequencies of the haplotypes are equal to the products of the frequencies of the alleles involved, and the equilibrium frequencies of the complex genotypes are equal to the products of the Hardy–Weinberg frequencies of the single-locus genotypes for the loci involved. The strength of the linkage between the loci is irrelevant with regard to the genotypic composition in the equilibrium. It only affects the number of generations of panmictic reproduction required to 'attain' the equilibrium.

Table. Presents the equilibrium frequencies of complex genotypes and phenotypes for the simultaneously considered loci *A-a* and *B-b*.

(*a*) *Genotypes*	*bb*	*Bb*	*BB*	
aa	q^2s^2	$2q^2rs$	q^2r^2	q^2
Aa	$2pqs^2$	$4pqrs$	$2pqr^2$	$2pq$
AA	p^2s^2	$2p^2rs$	p^2r^2	p^2
	s^2	$2rs$	r^2	1

(*b*) *Phenotypes*	*bb*	*B*	
aa	q^2s^2	$q^2(1-s^2)$	q^2
A	$(1-q^2)s^2$	$(1-q^2)(1-s^2)$	$(1-q^2)$
	s^2	$1-s^2$	

The foregoing is illustrated in Example, which deals with the production of a single-cross hybrid variety and the population resulting from its offspring as obtained by panmictic reproduction. Example illustrates the production of a synthetic variety and a few of its offspring generations as obtained by continued random mating.On the basis of the frequencies of the phenotypes for two traits (each with two levels of expression) showing qualitative variation, one can easily determine whether or not a certain population is in linkage equilibrium. It is, however, impossible to conclude whether or not the loci involved are linked. Only test crosses between individual plants with the phenotype $A \cdot B\cdot$ and plants with genotype *aabb* will give evidence about this. *N.B.* By 'phenotype $A \cdot B\cdot$' is meant the phenotype due to genotype *AABB*, *AaBB*, *AABb* or *AaBb*

Genotypic Composition	G_1 *for* G_0 *in C – phase*	G_1 *for* G_0 *in R – Phase*	$G\infty$
Genotype	$\frac{1}{4}(1-r_c)^2$	$\frac{1}{4}r_c^2$	$\frac{1}{16}$
aabb	$\frac{1}{2}r_c(1-r_c)$	$\frac{1}{2}r_c(1-r_c)$	$\frac{2}{16}$
aaBb	$\frac{1}{4}r_c^2$	$\frac{1}{4}(1-r_c)^2$	$\frac{1}{16}$
aaBB	$\frac{1}{2}r_c(1-r_c)$	$\frac{1}{2}r_c(1-r_c)$	$\frac{2}{16}$
Aabb	$\frac{1}{4}r_c^2$	$\frac{1}{2}r_c^2$	$\frac{2}{16}$
AB / ab	$\frac{1}{2}r_c(1-r_c)$	$\frac{1}{2}(1-r_c)^2$	$\frac{2}{16}$
Ab / aB	$\frac{1}{2}(1-r_c)^2$	$\frac{1}{2}rc(1-r_c)^2$	$\frac{2}{16}$
AaBB	$\frac{1}{4}r_c^2$	$\frac{1}{4}(1-r_c)^2$	$\frac{1}{16}$
AAbb	$\frac{1}{2}r_c(1-r_c)$	$\frac{1}{2}rc(1-r_c)$	$\frac{2}{16}$
AABb	$\frac{1}{4}(1-r_c)^2$	$\frac{1}{4}r_c^2$	$\frac{1}{16}$
AABB			

Example A synthetic variety is planned to be produced by intermating five clones of a self-incompatible grass species. Because crosses within each of the five components are excluded, the synthetic variety is produced by outbreeding. It is, therefore, due to a complex bulk cross. The obtained plant material is designated as Syn1 (or G0 in the present context). The five clones have the following genotypes for the two unlinked loci *B*1-*b*1 and *B*2-*b*2: clone 1: *b*1*b*1*b*2*b*2; clones 2 and 3: *B*1*B*1*b*2*b*2, and clones 4 and 5: *B*1*B*1*B*2*B*2.

The genotypic composition of Syn1 can be derived from the following scheme:

	$b_1b_1b_2b_2$	$B_1B_1b_2b_2$	$B_1B_1b_2b_2$	$B_1B_1B_2B_2$	$B_1B_1B_2B_2$
$b_1b_1b_2b_2$		$B_1b_1b_2b_2$	$B_1b_1b_2b_2$	$B_1b_1B_2b_2$	$B_1b_1B_2b_2$
$B_1B_1b_2b_2$	$B_1b_1b_2b_2$	–	$B_1B_1b_2b_2$	$B_1B_1B_2b_2$	$B_1B_1B_2b_2$
$B_1B_1b_2b_2$	$B_1b_1b_2b_2$	$B_1B_1b_2b_2$		$B_1B_1B_2b_2$	$B_1B_1B_2b_2$
$B_1B_1B_2B_2$	$B_1b_1B_2b_2$	$B_1B_1b_2b_2$	$B_1B_1B_2b_2$		$B_1B_1B_2B_2$
$B_1B_1B_2B_2$	$B_1b_1B_2b_2$	$B_1B_1b_2b_2$	$B_1B_1B_2b_2$	$B_1B_1B_2B_2$	

Table presents the genotype frequencies in a few relevant generations. When deriving these it was assumed that incompatibility can be neglected when considering continued panmictic reproduction starting in G0. The portion of homozygous plants in G0, G1, G2 and G– amounts to 0.2; 0.35; 0.3508 and 0.3536, respectively. The excess of heterozygous plants in comparison to the linkage equilibrium amounts therefore to 0.1536; 0.0036 and 0.0028 in G0, G1 and G2, respectively.

(This concerns plants which are heterozygous for one or two loci. For each single locus the Hardy–Weinberg genotypic composition occurs in G1 and all later generations).

The genotypic composition of plant material obtained when creating and maintaining an imaginary synthetic variety.

P indicates the parental clones, Go indicates population $syn_{1,}$ G_1 indicates $syn_{2,}$ G_2 indicates syn_3 and G– indicates syn–

Frequency

Genotype	*P*	*G*$_0$	*G*$_1$	*G*$_2$	*G*∞
$b_1b_1b_2b_2$	0.2		0.0225	0.0182	0.0144
$b_1b_1B_2B_2$			0.0150	0.0176	0.0192
$b_1b_1B_2B_2$			0.0025	0.0042	0.0064
$B_1b_1b_2b_2$		0.2	0.1350	0.1256	0.1152
$B_1B_2\,/\,b_1b_2$		0.2	0.1050	0.0904	0.0768
$B_1b_2\,/\,b_1B_2$			0.0450	0.0605	0.0768
$B_1b_1B_2B_2$			0.0350	0.0436	0.0512
$B_1b_1b_2b_2$	0.4	0.1	0.2025	0.2162	0.2304
$B_1B_1B_2b_2$		0.4	0.3150	0.3116	0.3072
$B_1B_1B_2B_2$	0.4	0.1	0.1225	0.1122	0.1024

The Haplotype Frequencies in Generation *t*

In this appendix, first is derived an equation relating the frequency of gametes with haplotype ab in generation t + 1 to its frequency in generation t, *i.e.* Equation.

Thereafter an equation describing the haplotype frequencies in generations due to continued panmictic reproduction, starting with a singlecross hybrid, is derived.

The Frequency of Gametes with Haplotype ab

The frequency of gametes with haplotype *ab*, produced by generation G*t*, are equal to

$$g00,t+1 = f00,t+\frac{1}{2}f10,t+\frac{1}{2}01,t+\frac{1}{2}(1-r_c)f11c,t+\frac{1}{2}r_cf11R,t$$

$$= f00,t+\frac{1}{2}f10,t+\frac{1}{2}f01,t+\frac{1}{2}f11c,t-r_cd_t$$

Panmictic reproduction of generation G*t* yields generation G*t* + 1. The genotypic composition of G*t* + 1 is described by the frequencies given by the third column of the previous table. Inclusion of these genotype frequencies in the following equations:The relevant genotypes, their frequencies (in general, as well as after panmixis) and the haplotypic composition of the gametes they produce are:

Genotype	*in general*	*After Panmixis*	*ab*	*aB*	*Ab*	*AB*
aabb	$f00$	$g00^2$	1	0	0	0
Aabb	$f10$	$2g00g10$	$\frac{1}{2}$	0	$\frac{1}{2}$	0
AAbb	$f20$	$g10^2$	0	0	1	0
aaBb	$f01$	$2g00g01$	$\frac{1}{2}$	$\frac{1}{2}$	0	0
$\frac{AB}{aB}$	$f11C$	$2g00g11$	$-\frac{1}{2}r_c$	$\frac{1}{2}r_c$	$\frac{1}{2}r_c$	$\frac{1}{2}$
$\frac{AB}{aB}$	$f11R$	$2g10g01$	$\frac{1}{2}r_c$	$\frac{1}{2}$	$\frac{1}{2}$	$-\frac{1}{2}r_c$
AABb	$f21$	$2g01g11$	0	$-\frac{1}{2}r_c$	$-\frac{1}{2}r_c$	$\frac{1}{2}r_c$
aaBB	$f02$	$g01^2$	0	0	$\frac{1}{2}$	$\frac{1}{2}$
AaBB	$f12$	$2g01g11$	0	1	0	0
AABB	$f22$	$g11^2$	0	$\frac{1}{2}$	0 0	$\frac{1}{2}$

Above equation for $g_{00,t+1}$ gives

$$g00,t+1 = g^2 00,t + g00,tg10,t + g00,tg01,t + g00,tg11,t - r_c d_t$$
$$= goo,t(g00,t + g10,t + g01,t + g11,t) - r_c d_t = g00,t - r_c d_t$$

Where, according to Equation

$d_t = (g11,tg00,t - g10,tg01,t)$

Similarly one can derive

$$g01,t+1 = g01,t + r_c d_t$$
$$g10,t+1 = g10,t + r_c d_t$$
$$g11,t+1 = g11,t - r_c d_t$$

The haplotype frequencies in generations due to continued panmictic reproduction, starting with a single-cross hybrid

In the case of panmictic reproduction starting from a single-cross hybrid there will be a symmetry in the haplotype frequencies such that

$$\text{g00, t} = \text{g11,t}$$

and

$$\text{g01, t} = \text{g10, t} = \frac{1}{2} - g11,t$$

Derivation of $g11,t$ suffices then to obtain the frequencies of all haplotypes with regard to two segregating loci. An equation presenting $g11,t$ immediately for any value for t will now be derived.

If the genotype of the single-cross hybrid is *AB/ ab, i.e.* coupling phase, the genotypic composition of the initial population G0 is simply described by

1. $f11C,0 = 1$, if it is *Ab/aB* the genotypic composition of G0 is described by $f11R,0 =$ Equation yields then

$$d_0 = \frac{1}{2}$$

in *the former case, and*

$$d_0 = \frac{-1}{2}$$

in the latter case. The frequency of gametes with the *AB* haplotype among the gametes produced by the single-cross amounts to

$$g11,1 = \frac{1}{2}(1 - r_c)$$

and

$$g11,1 = \frac{1}{2}r_c$$

Re*spectively in Example it was also derived that*

$$d_1 = \frac{1}{4}(1 - 2r_c)$$

for G_0 *in C – Phase and that*

$$d_1 = \frac{-1}{4(1 - 2r_c)}$$

respectively. In Example it was also derived that
for G0 in R-phase.

The frequencies of *AB* haplotypes in the case of continued panmixis follow from Equation combined with Equation:

$$\begin{aligned} g11,t+2 &= g11,t+1 - r_c d_t + 1 = g11,t+1 - r_c(1 - r_c)^t d_1 \\ &= g11,t - rc(1 - rc)^{t-1} d_1 - r_c(1 - rc)^t d1 \\ &= g11,1 - r_c d_1[(1 - r_c)^0 + \ldots + (1 - r_c)^t] \end{aligned}$$

The terms within the brackets form a convergent geometric series. The sum of such terms is given by the expression

$$a\frac{1 - q^n}{1 - q}$$

Where a is the first term, q is the multiplying factor and n is the number of terms. In the present situation this sum amounts to

$$\frac{1-(1-r_c)^{t+1}}{r_c}$$

$$\text{Thus } g11,t+2 = g11,1 - d_1[1-(1-r_c)^{t+1}]$$

$$For\ r_c = \frac{1}{2}\ we\ got\ d_1 = 0.\ Then$$

$$g11,t+2 = g1,1 = \frac{1}{4}$$

This implies that linkage equilibrium is present after one generation with panmictic reproduction!

For G0 in C-phase, Equation can be rewritten as

$$g11,t+2 = \frac{1}{2}(1-r_c) - \frac{1}{4}(1-2r_c)[1-(1-r_c)^{t+1}]$$

Thus

$$g11,2 = \frac{1}{2}(1-r_c) - \frac{1}{4}r_c(1-2r_c) = \frac{1}{2r}r_c^2 - \frac{3}{4}r_c + \frac{1}{2}$$

For G_0 in R-Phase, Equation can be transformed into

$$g11,t+2 = \frac{1}{2}r_c + \frac{1}{4}(1-2r_c)[1-(1-r_c)^{t+1}]$$

This implies

$$g11,2 = \frac{1}{2}r_c + \frac{1}{4}r_c(1-2r_c) = -\frac{1}{2}r_c^2 + \frac{3}{4}r_c$$

$$g11,3 = \frac{1}{2}r_c + \frac{1}{4}(1-2r_c)[1-(1-r_c)^2] = \frac{1}{2}r_c^3 - 1\frac{1}{4}r_c^2 + r_c$$

These equations are of relevance with regard to the question of whether it is advantageous, when it is aimed to promote the frequency of plants with a genotype due to recombination, to apply random mating in an F2 population of a self-fertilizing crop.

ONE LOCUS WITH MORE THAN TWO ALLELES

Multiple allelism does not occur in the populations considered so far. However, multiple allelism is known to occur in self- and cross-fertilizing crops. It may further be expected in three-way-cross hybrids, and their offspring, as well as in mixtures of pure lines (landraces or multiline varieties).

Example The intensity of the anthocyanin colouration in lettuce (*Lactuca sativa*), a self-fertilizing crop, is controlled by at least three alleles. The colour and location of the white leaf spots of white clover (*Trifolium repens*), a cross-fertilizing crop, are controlled by a multiple allelic locus. The expression for these traits appears to be controlled by a locus with at least 11 alleles. Another locus, with at least four alleles, controls the red leaf spots. (White clover is an autotetraploid crop with a gametophytic incompatibility system and a diploid chromosome behaviour; $2n = 4x = 32$).

The frequencies (f) of the genotypes $AiAj$ (with $i \leq j$; $j = 1,..., n$) for the multiple allelic locus $A1$-$A2$-... -An attain their equilibrium values following a single round of panmictic reproduction. The genotypic composition is then: f

Genotype		
A_1A_1...	$AjAj$...	A_nA_n
$p1^2$	$2pipj$	pn^2

The Proportion of homozygous plants is minimal for $pj = \frac{1}{n}(for\ j = 1,...,n)$

and amounts then to $n\left(\frac{1}{n}\right)^2 = \frac{1}{n}$;

ONE LOCUS WITH TWO ALLELES

The majority of situations considered in this book involve a locus represented by not more than two alleles. This is certainly the case in diploid species in the following populations:

- Populations tracing back to a cross between two pure lines, say, a single cross
- Populations obtained by (repeated) backcrossing (if, indeed, both the donor and the recipient have a homozygous genotype)

It is possibly the case in populations tracing back to a three-way cross or a double cross. It is improbable in other populations, like populations of cross-fertilizing crops, populations tracing back to a complex cross, landraces, multiline varieties. To keep (polygenic) models simple, it will often be assumed that each of the considered loci is represented by only two alleles. Quite often this simplification will violate reality. The situation of multiple allelic loci is explicitly considered. If the expression for the trait of interest is controlled by a locus with two alleles A and a (say locus A-a) then the probability distribution of the genotypes occurring in the considered population is often described by Probability

Genotype		
aa	Aa	AA
f_0	f_1	f_2

One may represent the probability distribution (in this book mostly the term genotypic composition will be used) by the row vector (f_0, f_1, f_2). The symbol f_j represents the probability that a random plant contains j A-alleles in its genotype for locus A-a, where j may be equal to 0, 1 or 2.

It has become custom to use the word genotype frequency to indicate the probability of a certain genotype and for that reason the symbol f is used.

The plants of the described population produce gametes which have either haplotype a or haplotype A. (Throughout this book the term haplotype is used to indicate the genotype of a gamete.) The probability distribution of the haplotypes of the gametes produced by the population is described by

Probability	Haplotype	
	a	A
	g_o	g_1

The symbol g_j represents the probability that a random gamete contains j Aalleles in its haplotype for locus A-a, where j may be equal to 0 or 1. The row vector (g_0, g_1) describes, in a condensed way, the haplotypic composition of the gametes. The habit to use the symbol q instead g_0 and the symbol p instead of g_1 is followed in this book whenever a single locus is considered. The term allele frequency will be used to indicate the probability of the considered allele.

So far it has been assumed that the allele frequencies are known and hereafter the theory is further developed without considering the question of how one arrives at such knowledge. In fact allele frequencies are often unknown.

When one would like to estimate them one might do that in the following way. Assume that a random sample of N plants is comprised of the following numbers of plants of the various genotypes: Number of Plants.

Genotype		
aa	Aa	AA
n_o	n_1	n_2

For any value for N the frequencies q and p of alleles a and A may then be estimated as

$$q = \frac{2n_0 + n_1}{2N} \text{ and } p = \frac{n_1 + 2n_2}{2N}$$

Throughout the book the expressions 'the probability that a random plant has genotype Aa', or 'the probability of genotype Aa', or 'the frequency of genotype Aa' are used as equivalents. This applies likewise for the expressions 'the probability that a gamete has haplotype A', or 'the probability of A'. Fusion of a random female gamete with a random male gamete yields a genotype specified by j, the number of A alleles in the genotype.

(The number of a alleles in the genotype amounts – of course – to $2 - j$.) The probability that a plant with genotype aa results from the fusion is in fact equal to the probability of the event that j assumes the value 0. The quantity j

assumes thus a certain value (0 or 1 or 2) with a certain probability. This means that *j* is a random variable.The probability distribution for *j*, *i.e.* for the genotype frequencies, is given by the binomial probability distribution:

$$P(j=j)=\binom{2}{j}p^j q^{2-j}$$

Fusion of two random gametes therefore yields

- With probability $q2$ a plant with genotype *aa*
- With probability $2pq$ a plant with genotype *Aa*
- With probability $p2$ a plant with genotype *AA*

The probabilities for the multinomial probability distribution of plants with these genotypes may be represented in a condensed form by the row vector ($q2$, $2pq$, $p2$). This notation represents also the genotypic composition to be expected for the population obtained after panmixis in a population with gene frequencies (q, p).In the case of panmixis there is a direct relationship between the gene frequencies in a certain generation and the genotypic composition of the next generation.

Thus if the genotype frequencies f_0, f_1 and f_2 of a certain population are equal to, respectively, q_2, $2pq$ and p_2, the considered population has the so-called Hardy–Weinberg (genotypic) composition. The actual genotypic composition is then equal to the composition expected after panmixis.

With continued panmixis, populations of later generations will continue to have the Hardy–Weinberg composition. Therefore such composition may be indicated as the Hardy–Weinberg equilibrium. The names of Hardy (1908) and Weinberg (1908) are associated with this genotypic composition, but it was in fact derived by Castle in 1903. With two alleles per locus the maximum frequency of plants with the *Aa* genotype in a population originating from panmixis $\frac{1}{2}$ *for* $p=q=\frac{1}{2}$. This occurs in F_2 populations of self-fertilizing crops. The F_2 originates from selfing of individual plants of the F_1, but because each plant of the

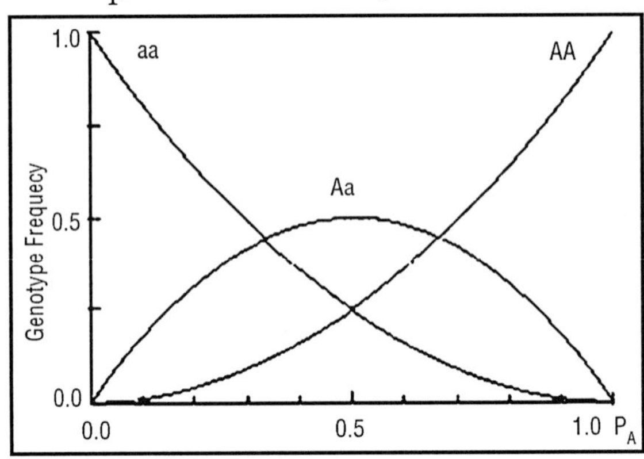

F_1 has the same genotype, panmixis within each plant coincides with panmixis of the F_1 as a whole. (The F1 itself may be due to bulk crossing of two pure lines; the proportion of heterozygous plants amounts then to 1.) The Hardy–Weinberg genotypic composition constitutes the basis for the development of population genetic theory for cross-fertilizing crops. It is obtained by an infinitely large number of pairwise fusions of random eggs with random pollen, as well as by an infinitely large number of crosses involving pairs of random plants. One may also say that it is expected to occur both after pairwise fusions of random eggs and pollen, and when crossing plants at random.

In a number of situations two populations are crossed as bulks. One may call this bulk crossing. One population contributes the female gametes (containing the eggs) and the other population the male gametes (the pollen, containing generative nuclei in the pollen tubes). In such a case, crosses within each of the involved populations do not occur. A possibly unexpected case of bulk crossing is described in Note.

Note: Selection among plants after pollen distribution, *e.g.* selection with regard to the colour of the fruits (if fruit colour is maternally determined), implies a special form of bulk crossing: the rejected plants are then excluded as effective producers of eggs (these plants will not be harvested), whereas all plants (could) have been effective as producers of pollen. The results, to be derived hereafter, in the main text, for a bulk cross of two populations with different allele frequencies.

A bulk cross is of particular interest if the haplotypic composition of the eggs differs from the haplotypic composition of the pollen. Thus if population I, with allele frequencies (q_1, p_1), contributes the eggs and population II, with allele frequencies (q_2, p_2), the pollen, then the expected genotypic composition of the obtained hybrid population, in row vector notation, is

$$(q_1q_2, p_1q_2 + p_2q_1, p_1p_2)$$

This hybrid population does not result from panmixis. The frequency of allele A is

$$P = \frac{1}{2}(p_1q_2 + p_2q_1) + p_1p_2 = \frac{1}{2}p_1q_2 + \frac{1}{2}p_1p_2 + \frac{1}{2}p_2q_1 + p_1p_2$$
$$= \frac{1}{2}p_1(q_2 + p_2) + \frac{1}{2}p_2(q_1 + p_1) = \frac{1}{2}(p_1 + p_2)$$

$$q_2 + p_2 = q_1 + p_1 = 1$$

Further equations based on $p + q = 1$ are elaborated When deriving Equation the equation $p + q = 1$ was used. On the basis of the latter equation several other equations, applied throughout this book, can be derived:

$$q^2 + 2pq + p^2 = 1$$
$$p - q = 2p - 1 = 1 - 2q$$
$$(p - q)^2 = p^2 - 2pq + q^2) = 1 - 4pq$$
$$p^2 - q^2 = (p + q)(p - q) = p - q = f_2 - f_0$$
$$p - q + 2pq = p^2 - q^2 + 2pq = p^2 + 2pq - q^2 = 1 - 2q^2$$

Panmictic reproduction of this hybrid population produces offspring with the Hardy–Weinberg genotypic composition.

The hybrid population contains, compared to the offspring population, an excess of heterozygous plants. The excess is calculated as the difference in the frequencies of heterozygous plants:

$$(p_1q_2 + p_2q_1) - 2pq = (p_1q_2 + p_2q_1) - 2[\frac{1}{2}(p_1 + p_2)\frac{1}{2}(q_1 + q_2)$$
$$= \frac{1}{2}(p_1q_2 + p_2q_1 - p_1q_1 - p_2q_2)$$
$$= \frac{1}{2}(p_1 - p_2)(q_2 - q_1) = \frac{1}{2}(p_1 - p_2)^2$$

This square is positive, unless $p_1 = p_2$. Thus the hybrid does indeed contain an excess of heterozygous plants.

Example illustrates that the superiority of hybrid varieties might (partly) be due to this excess. Example pays attention to the case of both inter- and intra-mating of two populations.

CONCLUSION

There are several mechanisms promoting cross-pollination and, consequently, cross-fertilization. The most important ones are

- Dioecy, *i.e.* male and female gametes are produced by different plants.
  - Asparagus *Asparagus officinalis* L.
  - Spinach *Spinacia oleracea* L.
  - Papaya *Carica papaya* L.
  - Pistachio *Pistacia vera* L.
  - Date palm *Phoenix dactylifera* L.
- Monoecy, *i.e.* male and female gametes are produced by separate flowers occurring on the same plant.
  - Banana *Musa* spp.
  - Oil palm *Elaeis guineensis* Jacq.
  - Fig *Ficus carica* L.
  - Coconut *Cocos nucifera* L.
  - Maize *Zea mays* L.

Cucumber *Cucumis sativus* L.

In musk melon (*Cucumis melo* L.) most varieties show andromonoecy, *i.e.* the plants produce both staminate flowers and bisexual flowers, whereas other varieties are monoecious.

- Protandry, *i.e.* the pollen is released before receptiveness of the stigmata.
  - Leek *Allium porrum* L.
  - Onion *Allium cepa* L.
  - Carrot *Daucus carota* L.
  - Sisal *Agave sisalana* Perr.
- Protogyny, *i.e.* the stigmata are receptive before the pollen is released.
  - Tea *Camellia sinensis* (L.) O. Kuntze
  - Avocado *Persea americana* Miller
  - Walnut *Juglans nigra* L.
  - Pearl millet *Pennisetum typhoides* L. C. Rich.
- Self-incompatibility, *i.e.* a physiological barrier preventing normal pollen grains fertilizing eggs produced by the same plant.
  - Cacao *Theobroma cacao* L.
  - Citrus *Citrus* spp.
  - Tea *Camellia sinensis* L. O. Kuntze
  - Robusta coffee *Coffea canephora* Pierre ex Froehner
  - Sugar beets *Beta vulgaris* L.
  - Cabbage, kale *Brassica oleracea* spp.
  - Rye *Secale cereale* L.
  - Many grass species, *e.g.* perennial ryegrass (*Lolium perenne* L.)
- Flower morphology
  - Fig *Ficus carica* L.
  - Primrose *Primula veris* L.
  - Common buckwheat *Fagopyrum esculentum* Moench.

and probably in the Bird of Paradise flower *Strelitzia reginae* Banks Effects with regard to the haplotypic and genotypic composition of a population due to (continued) reproduction by means of panmixis will now be derived for a so-called panmictic population. Panmictic reproduction occurs if each of the next five conditions apply:

- Random mating
- Absence of random variation of allele frequencies
- Absence of selection
- Absence of mutation
- Absence of immigration of plants or pollen In the remainder of this section the first two features of panmixis are more closely considered.

RANDOM MATING

Random mating is defined as follows: in the case of random mating the fusion of gametes, produced by the population as a whole, is at random with regard to the considered trait. It does not matter whether the mating occurs by means of crosses between pairs of plants combined at random, or by means of open pollination.

Open pollination in a population of a cross-fertilizing (allogamous) crop may imply random mating. This depends on the trait being considered. One should thus be careful when considering the mating system. This is illustrated in Example.

Example. Two types of rye plants can be distinguished with regard to their epidermis: plants with and plants without a waxy layer. It seems justifiable to assume random mating with regard to this trait. With regard to time of flowering, however, the assumption of random mating may be incorrect. Early flowering plants will predominantly mate *inter se* and hardly ever with late flowering plants. Likewise late flowering plants will tend to mate with late flowering plants and hardly ever with early flowering ones. With regard to this trait, so-called assortative mating occurs.

One should, however, realize that the ears of an individual rye plant are produced successively. The assortative mating with regard to flowering date may thus be far from perfect. Also, with regard to traits controlled by loci linked to the locus (or loci) controlling incompatibility, *e.g.* in rye or in meadow fescue (*Festuca pratensis*), perfect random mating will therefore probably not occur. Selection may interfere with the mating system. Plants that are resistant to an agent (*e.g.* disease or chemical) will mate *inter se* (because susceptible plants are eliminated). Then assortative mating occurs due to selection.

Crossing of neighbouring plants implies random mating if the plants reached their positions at random; crossing of contiguous inflorescences belonging to the same plant (geitonogamy) is, of course, a form of selfing.

Random mating does not exclude a fortuitous relationship of mating plants. Such relationships will occur more often with a smaller population size. If a population consists, generation after generation, of a small number of plants, it is inevitable that related plants will mate, even when the population is maintained by random mating.

Indeed, mating of related plants yields an increase in the frequency of homozygous plants, but in this situation the increase in the frequency of homozygous plants is also due to another cause: fixation occurs because of non-negligible random variation of allele frequencies. Both causes of the increase in homozygosity are due to the small population size (and not to the mode of reproduction).

This ambiguous situation, so far considered for a single population, occurs particularly when numerous small subpopulations form together a large superpopulation. In each subpopulation random mating, associated with non-negligible random variation of the allele frequencies, may occur, whereas in the superpopulation as a whole inbreeding occurs. Example provides an illustration. *Example.* A large population of a self-fertilizing crop, *e.g.* an F2 or an F3 population, consists of numerous subpopulations each consisting of a single plant. Because the gametes fuse at random with regard to any trait, one may state that random mating occurs within each subpopulation.

At the level of the superpopulation, however, selfing occurs. Selfing is impossible in dioecious crops, *e.g.* spinach (*Spinacia oleracea*). Inbreeding by means of continued sister × brother crossing may then be applied. This full sib mating at the level of the superpopulation may imply random mating within subpopulations consisting of full sib families.

Seen from the level of the superpopulation, inbreeding occurs if related plants mate preferentially. This may imply the presence of subpopulations, reproducing by means of random mating.

If very large, the superpopulation will retain all alleles. The increasing homozygosity rests on gene fixation in the subpopulations. If, however, only a single full sib family produces offspring by means of open pollination, implying crossing of related plants, then the population as a whole (in this case just a single full sib family) is still said to be maintained by random mating.

Absence of Random Variation of Allele Frequencies

The second characteristic of panmixis is absence of random variation of allele frequencies from one generation to the next. This requires an infinite effective size of the population, originating from an infinitely large sample of gametes produced by the present generation.

Panmixis thus implies a deterministic model. In populations consisting of a limited number of plants, the allele frequencies vary randomly from one generation to the next. Models describing such populations are stochastic models.

4

Effects of Inbreeding

SELF-FERTILIZATION AND CROSS-FERTILIZATION

There are many crops which are neither completely autogamous nor allogamous:

Broad bean *Vicia faba* L.
Oil-seed rape *Brassica napus* L.
Lupin *Lupinus luteus* L.
Sorghum *Sorghum bicolor* (L.) Moench.
Cotton *Gossypium hirsutum* L.
Safflower *Carthamus tinctorius* L.

The genotypic composition resulting from this mixture of modes of reproduction is considered. The portion of the eggs which develops into a zygote after selfing is represented by s and the portion which develops into a zygote after cross-fertilization by $k = 1-s$.

A general description of the genotypic composition of the plants of generation t is

	Genotype		
	aa	Aa	AA
f	$q^2 + pqf_t$	$2pq(1-f_t)$	$P^2 + pqf_t$

The portion s=1-K of the plants in generation t+1 originates then from selfing. Its genotypic composition is

	Genotype		
	aa	Aa	AA
f	$q^2 + pqf_t + \frac{1}{2}pq(1-f_t)$	$pq(1-f_t)$	$p^2 + pqf_t + \frac{1}{2}pq(1-f_t)$

The portion K of the plants in generation t+1 originates from random mating. Its genotypic composition is

Genotype

	aa	Aa	AA
f	q^2	$2pq$	P^2

Among all offspring the frequency of plants with a heterozygous genotype is then

$$f_{1,t+1} = 2pq(1-f_{t+1}) = (1-K).pq(1-f_t) + K.2pq$$

$$1-f_{t+1} = \frac{1}{2}(1-K)(1-f_t)+K$$
$$2-2f_{t+1} = 1-K-f_t+Kf_t+2K$$
$$2f_{t+1} = 1-K+f_t-Kf_t = (1-K)(1+f_t)$$
$$f_{t+1} = \frac{1}{2}S(1+f_t)$$

As required, this expression coincides at $s = 1$ with Equation. We now consider the situation that s is constant from one generation to the next. In the case of equilibrium, successive generations have identical genotypic compositions. Then $Ft = Ft+1 = Ft+2 = \ldots = Fe$. Equation implies then

$$2f_e = S(1+f_e) = S+Sf_e$$
$$f_e(2-S) = S$$

Thus

$$f_e = \frac{S}{2-S}$$

In the equilibrium (e) the genotypic composition is

Genotype

	aa	Aa	AA
f	q^2+pqf_e	$2pq(1-f_e)$	p^2+pqf_e

The relation between Fe and s, *i.e.* Equation, is almost linear in the range of possible values for s: Fe roughly equals s. We now consider, for the case of $p = q = 1/2$, the effect on the genotypic composition of a continued change in the mode of reproduction. First the

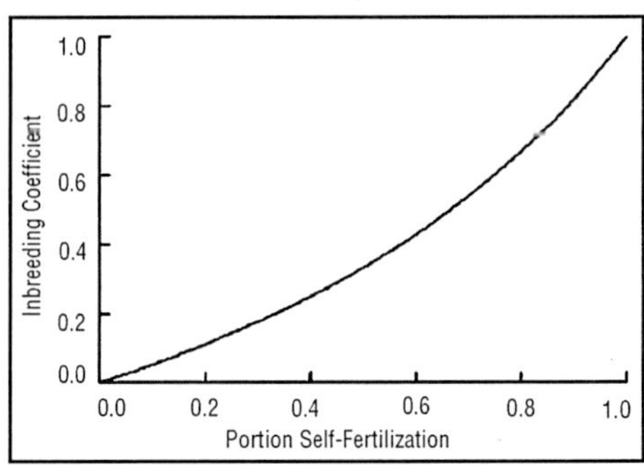

Population genetical effect of some cross-fertilization, *i.e.* $k > 0$, in an – until then - exclusively self-fertilizing crop (*e.g.* wheat) is considered; thereafter we consider the population genetical effect of some selfing, *i.e.* $s > 0$, in an – until then – exclusively cross-fertilizing crop.

Population genetical effect of some cross-fertilization, *i.e.* $k > 0$, in an – until then - exclusively self-fertilizing crop (*e.g.* wheat) is considered; thereafter we consider the population genetical effect of some selfing, *i.e.* $s > 0$, in an – until then – exclusively cross-fertilizing crop. Assume that in an $F\infty$population, with genotypic composition (1/ 2, 0, 1/ 2), from some generation onward always 10per cent of the offspring result from crossfertilization (*i.e.* $k = 0.1$), *e.g.* because the population is maintained in a different environment.

In this case the frequency of heterozygous plants increases from $f1 = 0$ to $f1,e = 0.09$. Some cross-fertilization in a self-fertilizing crop gives thus a non-negligible increase in the frequency of heterozygous plants. According to Equation the successive generations will have the following coefficients of inbreeding:

$$f_1 = 0.900$$

$$f_2 = 0.855$$

$$f_3 = 0.835$$

$$f_4 = 0.826$$

$$f_e = 0.818$$

It is concluded that equilibrium is approached slowly.

Some Self-Fertilization in a Cross-Fertilizing Crop

We consider a panmictic population with genotypic composition (1 4, 1 2, 1 4). From some generation onward always 10per cent of the offspring is due to selfing (*i.e.* $s = 0.1$). This results in a reduction of the frequency of heterozygous plants: at $s = 0.1$ it reduces from $f1 = 0.50$ to $f1,e = 0.47$. It can be derived that

$$f_1 = 0.050$$

$$f_2 = 0.053$$

$$f_e = 0.053$$

In this situation the equilibrium is attained almost immediately.

Workman and Allard (1962) studied the equilibrium with regard to two segregating loci, attained in the case of simultaneous occurrence of selfing and cross-fertilization, for unlinked loci. Weir and Cockerham (1973) did so for linked loci.

DIPLOID CHROMOSOME BEHAVIOUR AND INBREEDING

With continued inbreeding of any (infinitely) large population the genotype frequencies will change from one generation to the other until the frequency of plants with a heterozygous genotype has become zero.

Starting from the initial population G0 with genotypic composition ($f0, 0, f1, 0, f2, 0$), eventually a population with genotypic composition (q, 0, p) will be obtained.

Table The Frequency of Genotypes aa, Aa and AA in the Case of Continued Selfing

(a) Starting with some arbitrary genotypic composition

	Genotype		
Generation	*aa*	*Aa*	*AA*
$S0$	$f0$	$f1$	$f2$
$S1$	$f0+\frac{1}{4}f1$	$\frac{1}{2}f1$	$f2+\frac{1}{4}f1$
$S2$	$f0+\left(\frac{1}{4}+\frac{1}{8}\right)f1$	$\frac{1}{4}f1$	$f2+\left(\frac{1}{4}+\frac{1}{8}\right)f1$
$S3$	$f0+\left(\frac{1}{4}+\frac{1}{8}+\frac{1}{16}\right)f1$	$\frac{1}{8}f1$	$f2+\left(\frac{1}{4}+\frac{1}{8}+\frac{1}{16}\right)f1$
.			
.			
$S\infty$	q	0	P

(b) Starting with F1, *i.e.* a population with genotypic composition (0, 1, 0)

Generation		*Inbreeding*	*Panmictic*		*Genotype*	
(*t*)	*Population*	*Coefficient*($\wp$	*index*(*P*)	*aa*	*Aa*	*AA*
0	$S_0(=F_1)$	−1	2	0	1	0
1	$S_1(=F_2)$	0	1	$\frac{1}{4}$	$\frac{1}{2}$	$\frac{1}{4}$
2	$S_2(=F_3)$	$\frac{1}{2}$	$\frac{1}{2}$	$\frac{3}{8}$	$\frac{2}{8}$	$\frac{3}{8}$
3	$S_3(=F_4)$	$\frac{3}{4}$	$\frac{1}{4}$	$\frac{7}{16}$	$\frac{2}{16}$	$\frac{7}{16}$
4	$S_4(-F_5)$	$\frac{7}{8}$	$\frac{1}{8}$	$\frac{15}{32}$	$\frac{2}{32}$	$\frac{15}{32}$
5	$S_5(=F_6)$	$\frac{15}{16}$	$\frac{1}{16}$	$\frac{31}{64}$	$\frac{2}{64}$	$\frac{31}{64}$
6	$S_6(=F_7)$	$\frac{31}{32}$	$\frac{1}{32}$	$\frac{63}{128}$	$\frac{2}{128}$	$\frac{63}{128}$
7	$S_7(=F_8)$	$\frac{63}{64}$	$\frac{1}{64}$	$\frac{127}{256}$	$\frac{2}{256}$	$\frac{127}{256}$
∞	$S\infty(=F\infty)$	1	0	$\frac{1}{2}$	0	$\frac{1}{2}$

Illustrates this for inbreeding by means of continued selfing. It appears that the genotype frequencies approach, in an asymptotic manner, the gene and haplotype frequencies.Often the frequency of heterozygous plants in generation t, *i.e.* $f1,t$, is written in the form

$$2pq(1-f_t)$$

In this expression the factor 1–*Ft* describes the deviation from the Hardy–Weinberg frequency. The factor is called the panmictic index, sometimes designated by the symbol P.

This implies that $P = 1 - Ft$. The parameter *Ft*, say 'script F', is the inbreeding coefficient (or fixation index) pertaining to generation t.When starting with an F1 population, F2 is the first generation due to self-fertilization. For this reason the F2 population is chosen to be generation 1. (Its genotypic composition is equal to the genotypic composition of the population obtained by panmictic reproduction of the F1) Successive generations may be indicated by G1, G2,..., but in the case of continued selfing the designations S1, S2, S3,... are used as well. A general description of the genotypic composition of any population (inbred or not) is now given by

Genotype			
	aa	*Aa*	*AA*
f	$q^2 + pqf_t$	$2pq(1-f_t)$	$P^2 + pqf_t$

In several other books, *e.g.* Falconer and MacKay (1996), the inbreeding coefficient is defined as the probability that the two alleles at any loci of a plant are identical by descent. This would mean that the inbreeding coefficient of an F2 population obtained from cross *AA* × *aa* is equal to 1 2, because 50per cent of the plants contain, for locus *A-a*, alleles that are identical by descent (this concerns plants with genotype *aa* or *AA*). In this book the parameter F is used to quantify the deviations from the Hardy–Weinberg frequencies.

In an F2 population such deviations are absent and accordingly its inbreeding coefficient is 0. In Note, it is shown that our definition of the inbreeding coefficient F can be interpreted as the coefficient of correlation of numerical values, *e.g.* gene-effects, assigned to the haplotypes of the uniting gametes. This is based on the following consideration. With random mating the gene effects of the haplotypes of fusing female and male gametes are independent; in the absence of random mating they are interdependent.

With inbreeding they tend to be similar; with outbreeding they tend to be different. Breeding of self-fertilizing crops starts mostly with crossing of homozygous lines. For all loci for which the parental lines have a different homozygous genotype the genotype of the F1 is heterozygous. For these loci $p = q = 1\ 2$ and then the expressions in (3.1) simplify to

Genotype			
	aa	*Aa*	*AA*
f	$\frac{1}{4}(1+f_t)$	$\frac{1}{2}(1-f_t)$	$\frac{1}{4}(1+f_t)$

As $f1, 0 = 1\ 2\ (1–F0) = 1$, it follows that $F0 = -1$, *i.e.* a negative value for the inbreeding coefficient. The panmictic index of the F1 amounts for heterozygous loci to $P0 = 2$.

In the remainder of this section the decrease in the frequency of heterozygous plants is considered for the three most important regular inbreeding systems, *viz.* self-fertilization, full sib mating and parent × offspring mating. To measure this decrease the parameter λ is defined:

$$\lambda = \frac{2pq(1-f_t)}{2pq(1-f_{t-1})} = \frac{1-f_t}{1-f_{t-1}}$$

This parameter indicates the frequency of heterozygous plants as a proportion of this frequency in the preceding generation. At a smaller value for λ the decrease of $f1$ is stronger. In the case of selfing the values for λ do not depend on t; they are approximately constant when applying full sib mating or parent × offspring. Then $\lambda 1 = \lambda 2 = \cdots = \lambda t$. This implies

$$f1,t = \lambda f_{1,t-1} = \lambda^2 f1,t-2 = \lambda^t f1,0$$

In the F2 generation, the first generation generated by selfing, the genotype frequencies coincide with the Hardy-Weinberg frequencies. Thus $f1, 1 = 2pq$, implying that $F1$, the inbreeding coefficient of F2, is zero. In population F–, approximately obtained after a very large number of generations reproducing by means of selfing, there is complete homozygosity, *i.e.* $f1, - = 0$, implying that $F-$, the inbreeding coefficient of F–, is 1. The decrease of $f1$, due to continued selfing, is indicated in Table. The table shows that $f1$ is halved by each round of reproduction by means of selfing. Thus

$$1 - f_t = \frac{1}{2}(1 - f_{t-1})$$

implying

$$f_t = \frac{1}{2}(1 + f_{t-1})$$

With Regard to Continued selfing the expression

$$1-f_t=\frac{1}{2}(1-f_{t-1})$$

or

$$P_t=\frac{1}{2}P_{t-1}$$

implies

$$P_t=\left(\frac{1}{2}\right)^t P_0=\left(\frac{1}{2}\right)^{t-1}$$

$$f_t=1-\left(\frac{1}{2}\right)^{t-1}$$

At all other systems of inbreeding the reduction of $f1$ is smaller. The minimum value for l is thus attained with selfing. It amounts to lS = 1/ 2. Li (1976, pp. 312–317) showed that for both full sib mating and parent × offspring mating, the relation

$$f1,t+2=\frac{1}{2}f1,t+1+\frac{1}{4}f1,t$$

Applies. Consider an initial population with genotypic composition (0,1,0), thus $f1,0$ = 1. In this population plants are crossed in pairwise combinations. In the next generation the genotypic composition of the population obtained, which consists of full sib families, is expected to be (1/ 2, 1/ 4, 1/ 2), with $f1,1$ = 1 2. Continued full sib mating, within the continuously generated FS-families, gives, according to Equation

$$f1,2=\frac{1}{2}\left(\frac{1}{2}\right)+\frac{1}{4}(1)=\frac{1}{2}, i.e. \lambda_2=1$$

$$f1,3=\frac{1}{2}\left(\frac{1}{2}\right)+\frac{1}{4}\left(\frac{1}{2}\right)=\frac{3}{8}, i.e. \lambda_3=\frac{3}{4}=0.75$$

$$f1,4=\frac{1}{2}\left(\frac{3}{8}\right)+\frac{1}{4}\left(\frac{1}{2}\right)=\frac{5}{16}, i.e. \lambda_4=\frac{5}{6}=0.8333, etc$$

The first round of inbreeding (full sib mating or parent × offspring mating) does not give a decrease of the frequency of heterozygous plants (λ2 = 1). Indeed, with full sib mating first FS-families have to be generated.

It appears that λ approaches asymptotically the value λFS = λPO = 0.809. As (0.809)3 = 0.53 ≈ 1/ 2, three generations of reproduction by means of FSmating or parent × offspring mating give the same reduction in $f1$ as a single round of reproduction by selfing.

AUTOTETRAPLOID CHROMOSOME BEHAVIOUR AND SELF-FERTILIZATION

Spontaneous self-fertilization as the natural mode of reproduction occurs rather rarely among crops with an autotetraploid chromosome behaviour. The somatic chromosome number of quinoa (*Chenopodium quinoa*) is $2n = 36$. The basic chromosome number for the genus *Chenopodium* is $x = 9$. This suggests that quinoa is a tetraploid. Ward (2000) found for the same locus both diploid and tetraploid behaviour. Simmonds (1976) reported that selfing predominates, without evident inbreeding depression.

Quite a few autotetraploid crops, *e.g.* durum wheat (*Triticum durum*; $2n = 4x = 28$) or coffee (*Coffea arabica*; $2n = 4x = 44$), have a diploid chromosome behaviour. For other crops, *e.g.* European potato (*Solanum tuberosum*; $2n = 4x = 48$) or wild barley (*Hordeum bulbosum*; $2n = 4x = 28$), there may be a more or less perfect autotetraploid chromosome behaviour, implying that exclusively quadrivalents are being formed at meiosis. Artificial selffertilization may be applied in a man-made autotetraploid crop such as rye (*Secale cereale*; $2n = 4x = 28$), which is self-incompatible in its natural diploid condition.

In this section attention is only given to the simple situation of a single segregating locus with two alleles. It is assumed that double reduction does not occur.

The genotypic composition of some initial generation, say S0, is

Genotype

	aaaa	*Aaaa*	*AAaa*	*AAAa*	*AAAA*
	nulliplex	*Simplex*	*Duplex*	*triplex*	*quadruplex*
f	f_0	f_1	f_2	f_3	f_4

Its gene frequencies are

$$P = \frac{1}{4}f_1 + \frac{1}{2}f_2 + \frac{3}{4}f_3 + f_4$$

and

$$q = 1 - P$$

It is first verified that the gene frequencies remain constant from one generation to the next (such constancy is to be expected in the absence of selection). In order to do this, Table is used. This table presents, for each possible autotetraploid genotype, and according to the haplotype frequencies presented in Table, the genotypic composition of the line obtained by selfing.

The allele frequencies in the parental population follow from Equation. Across the total of the lines obtained from this parental population the frequency of allele A is

$$\frac{1}{4}\left(\frac{1}{2}f_1 + \frac{2}{9}f_2\right) + \frac{1}{2}\left(\frac{1}{4}f_1 + \frac{1}{2}f_2 + \frac{1}{4}f_3\right) + \frac{3}{4}\left(\frac{2}{9}f_2 + \frac{1}{2}f_3\right)$$

$$+\left(\frac{1}{36}f_2 + \frac{1}{4}f_3 + f4\right) = \frac{1}{4}f_1 + \frac{1}{2}f_2 + \frac{3}{4}f_3 + f_4$$

This is equal to the frequency in the parental population. The genotypic composition of S– will thus be:

	Genotype				
	aaaa	*Aaaa*	*AAaa*	*AAAa*	*AAAA*
f	q	0	0	0	P

How fast do the frequencies of plants with a heterozygous genotype and of gametes with a heterozygous haplotype decrease with (continued) selfing?

Table. The genotypic composition of the line obtained by selfing an autotetraploid genotype.

Parent		Genotypic composition of line				
genotype	f	*aaaa*	*Aaaa*	*AAaa*	*AAAa*	*AAAA*
aaaa	f_0	1	0	0	0	0
Aaaa	f_1	$\frac{1}{4}$	$\frac{1}{2}$	$\frac{1}{4}$	0	0
AAaa	f_2	$\frac{1}{36}$	$\frac{2}{9}$	$\frac{1}{2}$	$\frac{2}{9}$	$\frac{1}{36}$
AAAa	f_3	0	0	$\frac{1}{4}$	$\frac{1}{2}$	$\frac{1}{4}$
AAAA	f_4	0	0	0	0	1

In order to answer this question, first the decrease of $g1$, *i.e.* the frequency of gametes with haplotype *Aa* is considered and thereafter the decrease of fh. *i.e.* the frequency of heterozygous plants. From Table it can be derived that

$$g1,t+1=\frac{1}{2}f1,t+\frac{4}{6}f2,t+\frac{1}{2}f3,t$$

Thus similarly,

$$g_{1,t+2}=\frac{1}{2}f_{1,t+1}\frac{4}{6}f_{2,t+1}+\frac{1}{2}f_{3,t+1}$$

$$=\frac{1}{2}\left(\frac{1}{2}f_{1,t}+\frac{2}{9}f_{2,t}\right)+\frac{4}{6}\left(\frac{1}{4}f_{1,t}+\frac{1}{2}f_{2,t}+\frac{1}{4}f_{3,t}\right)+\frac{1}{2}\left(\frac{2}{9}f_{2,t}+\frac{1}{2}f_{3,t}\right)$$

$$=\frac{5}{12}f_{1,t}+\frac{5}{9}f_{2,t}+\frac{5}{12}f_{3,t}$$

$$=\frac{5}{6}g_{1,t+1}$$

This implies that each population obtained by selfing still produces 5 6 of the proportion of gametes with the *Aa* haplotype which was produced by the previous generation.Now the frequency of plants with a heterozygous genotype is considered. This frequency is designated by fh. Thus

$$f_{h,t} = f_{1,t} + f_{2,t} + f_{3,t}$$

$$f_{1,t+2} = \frac{1}{2} f_{1,t+1} + \frac{2}{9} f_{2,t+1}$$

$$f_{2,t+2} = \frac{1}{4} f_{1,t+1} + \frac{1}{2} f_{2,t+1} + \frac{1}{4} f_{3,t+1}$$

$$f_{3,t+2} = \frac{2}{9} f_{2,t+1} + \frac{1}{2} f_{3,t+1}$$

The decrease of f_h at (continued) selfing is described by:

$$f_{h,t+2} = \frac{3}{4} f_{1,t+1} + \frac{17}{18} f_{2,t+1} + \frac{3}{4} f_{3,t+1}$$

$$= f_{h,t+1} - \left(\frac{1}{4} f_{1,t+1} + \frac{1}{18} f_{2,t+2} + \frac{1}{4} f_{3,t+1} \right)$$

$$= f_{h,t+1} - \left[\frac{1}{4}\left(\frac{1}{2} f_{1,t} + \frac{2}{9} f_{2,t} \right) + \frac{1}{18}\left(\frac{1}{4} f_{1,t} + \frac{1}{2} f_{2,t} + \frac{1}{4} f_{3,t} \right) + \frac{1}{4}\left(\frac{2}{9} f_{2,t} + \frac{1}{2} f_{3,t} \right) \right]$$

$$= fh,t+1 - \frac{5}{36}(f_{1,t} + f_{2,t} + f_{3,t})$$

$$= fh,t+1 - \frac{5}{36} f_{h,t}$$

We consider the decrease of the frequency of heterozygous plants for an initial population consisting exclusively of duplex plants. The genotypic composition of S0 is then (0, 0, 1, 0, 0), with $fh,0 = 1$. According to Table, $fh,1$ amounts

Table: The frequency in generation t of plants with a heterozygous genotype, viz, $f_{h,t,}$ in the case of continued self-fertilization in an autotetraploid population, starting with a population exclusively consisting of duplex plants. The parameterls indicates the portion of heterozygous plants which remained

Generation	*t*	fh,t	$\lambda_S = \frac{f_{h,t}}{f_{h,t-1}}$
S_0	0	1	
S_1	1	$\frac{17}{18} = 0.9444$	0.9444
S_2	2	$\frac{29}{36} = 0.8056$	0.8529
S_3	3	$\frac{437}{648} = 0.6744$	0.8372
S_4	4	$\frac{729}{1296} = 0.5625$	0.8341

Then to 2/ 9 + 1/ 2 + 2/ 9 = 17/ 18. Table presents the frequency of plants with a heterozygous genotype in successive generations, as calculated from Equation.

The frequency of heterozygous plants as a proportion of the frequency in the preceding generation, *i.e.*

$$\lambda_S = \frac{f_{h,t}}{f_{h,t-1}}$$

is also presented in Table. It appears that λ*S* converges to a constant value, *viz.* to 5/ 6 = 0.8333. This implies, per round of reproduction by selfing, the same constant (relative) decrease in the frequency of heterozygous plants as derived from the frequency of heterozygous gametes. In this phase, reproduction by means of self-fertilization for *n* successive generations reduces *fh,t* to

$$f_{h,t+n} = \left(\frac{5}{6}\right)^n f_{h,t}$$

The frequency of heterozygous plants is halved if

$$\left(\frac{5}{6}\right)^n = 0.5, i.e. \ if$$

$$n = \frac{In(0.5)}{In(0.8333)} = 3.8$$

Starting with an initial population with genotypic composition (0, 0, 1, 0, 0) the decrease of the frequency of heterozygous plants is even less: in S4, *fh,* 4 is still larger than 1/ 2.When comparing the decrease in the frequency of plants with a heterozygous genotype occurring at selfing of a diploid crop and such decrease at selfing of a diploid crop and such decrease at selfing of a tetraploid crop it is clear that the decrease is quite slow in the case of tetraploidy. Continued FS-mating in a diploid crop gives a somewhat faster decrease in the frequency of heterozygous plants than continued selfing of a tetraploid crop.

A more comprehensive treatment of population genetical effects of selfing in an autotetraploid population is given by Seyffert.

TWO OR MORE UNLINKED LOCI, EACH WITH TWO ALLELES

Independent segregation occurs when the recombination value is equal to 1/ 2. Some population genetical implications of continued selfing with regard to unlinked loci are thus easily obtained from results derived in Section. Consider the haplotypes *ab, aB, Ab* or *AB* for the two unlinked loci *A-a* and *B-b*. Equation shows that absence of linkage implies constancy of the haplotype frequencies

$$g_{00,t} + 1 = g_{00,t}$$

$$g_{01,t} + 1 = g_{01,t}$$

$$g_{10,t} + 1 = g_{10,t}$$

$$g_{11,t} + 1 = g_{11,t}$$

This applies for any genotypic composition of the initial population. An application is described in Note. The haplotypic composition of the gametes produced by populations S0, S1,..., S– remains thus constant across the generations. This implies that the genotypic composition of S– immediately follows from the haplotypic composition of the gametes produced by S0:

Note: When breeding a *non-perennial* cross-fertilizing crop, selection among plants on the basis of a progeny test is impossible because the candidate plants cannot be maintained. In such cases these plants are selfed: their S1-lines produce gametes with the same haplotypic composition as they do themselves. Indeed: haplotypic compositions can be maintained by means of selfing. This is applied in *recurrent selection for general combining ability* as well as in *reciprocal recurrent selection.*

	Genotype			
	aabb	*aaBB*	*AAbb*	*AABB*
f	g_{00}	g_{01}	g_{10}	g_{11}

The constancy of the haplotypic composition in the case of continued selfing is in striking contrast to the continuous change, until linkage equilibrium is attained, of the haplotypic composition in the case of continued panmixis. Notwithstanding the stability of the haplotype frequencies the genotype frequencies change drastically: the frequencies of heterozygous plants decrease and those of homozygous plants increase. The frequencies of the complex genotypes only become stable if heterozygous plants no longer occur.

When starting with an F1 the frequencies of the complex genotypes follow directly from the frequencies of the single-locus genotypes given by Equation. (It should be realized that in cross-fertilizing crops this rule applies only

The frequencies of complex and single-Locus genotypes for the unlinked loci A-a and B-b in generation t(=1,2,3,....–) produced by selfing for t generations since the F_1 Population

Genotype for Locus B-b

	bb	Bb	BB	
aa	$\frac{1}{16}(1+f_t)^2$	$\frac{1}{8}(1-f_t)^2$	$\frac{1}{16}(1+f_t)^2$	$\frac{1}{4}(1+f_t)$
Aa	$\frac{1}{8}(1-f_t)^2$	$\frac{1}{4}(1-f_t)^2$	$\frac{1}{8}(1-f_t)^2$	$\frac{1}{2}(1-f_t)$
AA	$\frac{1}{16}(1+f_t)^2$	$\frac{1}{8}(1-f_t)^2$	$\frac{1}{16}(1+f_t)^2$	$\frac{1}{4}(1+f_t)$
	$\frac{1}{4}(1+f_t)$	$\frac{1}{2}(1-f_t)$	$\frac{1}{4}(1+f_t)$	1

In linkage equilibrium). Thus Table presents the genotypic composition with regard to the complex genotypes for two unlinked loci of any generation obtained by (continued) selfing starting with an F1. It is, in general, impossible to determine how many loci control the phenotypic expression of a certain trait, *e.g.* culm length in wheat. The reason for this is that the contribution due to non-segregating loci cannot be assessed: if one crosses some line P1 with genotype *AabbccDD* with regard to the trait under consideration with line P2 with genotype *aabbCCdd* then the contribution due to locus *B-b* cannot be assessed. Thus it might appear that three instead of four loci are responsible for the genetic control of the trait.

In fact only the number of segregating loci, *i.e.* the number of loci for which the two homozygous parents have a different genotype with regard to the trait under consideration, can be studied. This number is an interesting quantity, upon which the size of an F2 generation (or a later generation) may be based. It is speculated that the analysis of (quantitative trait) loci based on molecular markers is going to substitute biometrical methods for estimating the number of segregating loci. When generating a large number of molecular markers one can localize (and count) polygenes with relatively large phenotypic effects on the studied trait.

We consider, for the case of K unlinked loci, the probability that a plant contains for k of these loci a heterozygous single-locus genotype and for the remaining K–k loci a homozygous genotype. This probability is given by the binomial probability distribution function:

$$P(\underline{K}=K)=\binom{K}{k}.\left(\frac{1-f_t}{2}\right)^k\left(\frac{1+f_t}{2}\right)^{K-k}$$

The probability of a completely homozygous plant is

$$P(\underline{K}=0)=\left(\frac{1+f_t}{2}\right)^K$$

Table: The probability of a completely homozygous plant in generation G_t($t = 1,\ldots\ldots.7$), obtained after t successive generations with reproduction by means of selfing, when considering K= 1,...,14 unlinked loci. Gt corresponds to generation F_{t+1}

K	t 1	2	3	4	5	6	7
1	0.500	0.750	0.875	0.938	0.969	0.984	0.992
2	0.250	0.563	0.766	0.879	0.938	0.969	0.984
3	0.125	0.422	0.670	0.824	0.909	0.954	0.977
4	0.063	0.316	0.586	0.772	0.881	0.939	0.969
5	0.031	0.237	0.513	0.724	0.853	0.924	0.962
6	0.016	0.178	0.449	0.679	0.827	0.910	0.954
7	0.008	0.133	0.393	0.637	0.801	0.896	0.947
8	0.004	0.100	0.344	0.597	0.776	0.882	0.939
9	0.002	0.075	0.301	0.559	0.751	0.868	0.932
10	0.001	0.056	0.263	0.524	0.728	0.854	0.925
11	0.000	0.042	0.230	0.492	0.705	0.841	0.917
12	0.000	0.032	0.201	0.461	0.683	0.828	0.910
13	0.000	0.024	0.176	0.432	0.662	0.815	0.903
14	0.000	0.018	0.154	0.405	0.641	0.802	0.896

or, When applying Equation

$$\left[\frac{1+1-\left(\frac{1}{2}\right)^{t-1}}{2}\right]^K=\left[1-\left(\frac{1}{2}\right)^t\right]^K=\left(\frac{2^{t-1}}{2^t}\right)^K$$

Table: Presents this probability for K=1,...,14 and t = 1,...,7. Allard gives a graphical presentation of these probabilities. The expected value of K the number of loci with a heterozyhgous single-locus genotype in a random plant, is

$$E\underline{K}=K.\frac{1}{2}(1-f_t)=\frac{1}{2}K\left(\frac{1}{2}\right)^{t-1}=\left(\frac{1}{2}\right)tK$$

It is $\frac{1}{2}K$ *in an* F_2 *plant,* $\frac{1}{4}K$ *in an* F_3 *plant, etc.*

The variance of $\underline{K}$ *is*

$$Var(\underline{K})=K.\frac{1}{2}(1-f_t).\frac{1}{2}(1+f_t)$$

$$=\frac{1}{4}K(1-f_t^2)=\frac{1}{4}K\left[1-\{1-\left(\frac{1}{2}\right)^{t-1}\}^2\right]$$

$$=\frac{1}{4}K\left[1-\left\{1-\left(\frac{1}{2}\right)^t+\left(\frac{1}{4}\right)^{t-1}\right\}\right]=\left[\left(\frac{1}{2}\right)^{t-2}-\left(\frac{1}{4}\right)^t\right]K$$

A PAIR OF LINKED LOCI

It was shown that linkage may be expected to play a relatively unimportant role in the inheritance of quantitative traits. It was said that, throughout this book, absence of linkage would be assumed. It is, nevertheless, useful to be familiar with some implications of linkage. An important reason for this is the study of the linkage of loci affecting a quantitative trait with molecular markers.

Consider haplotypes *ab*, *aB*, *Ab or AB* for the two loci *A*-*a* and *B*-*b* with recombination value *rc*. Continued selfing, starting with an F1 with the heterozygous genotype *AaBb*, yields in the absence of selection

$$g11,t = g00,t$$

$$g01,t = g10,t$$

'symmetric' haplotype frequencies:

$$g11,t + g10,t = PA = \frac{1}{2}$$

$$g10,t = \frac{1}{2} - g11,t$$

This implies that, when one knows $g11,t$, one also knows $g10,t$, $g01,t$ and $g00,t$. It suffices thus to consider only the frequency of gametes with the *AB* haplotype. This is particularly of interest when considering F–. This population is described by

	Genotype			
	aabb	*AAbb*	*aaBB*	*AABB*
f	$f00,\infty$	$f20,\infty$	$f02,\infty$	$f22,\infty$

Only plants with the *AABB* genotype are capable of producing gametes with the *AB* haplotype. Thus $g11,- = f22,-$. The haplotypic composition of the gametes produced by this population is

	Haplotype			
	ab	*Ab*	*aB*	*AB*
g	$g00,\infty(= g11,\infty)$	$g10,\infty(= \frac{1}{2} - g11,\infty)$	$g01,\infty(\frac{1}{2} - g11,\infty)$	$g11,\infty$

There are thus good reasons to consider the frequency of gametes with the *AB* haplotype. In Note the following relation between the frequencies of *AB*-haplotypes in two successive generations is derived:

Note: The frequency of *AB* haplotypes, *i.e.* $g11$, is considered for the case of continued autogamous reproduction. (To promote readability the recombination value is – in this section – mostly just indicated by the symbol *r*). The genotypes capable of producing *AB* haplotypes, their frequencies in generation *t* and the haplotypic composition of the gametes they produce are Haplotype

Genotype	*f*	*ab*	*aB*	*Ab*	*AB*
AABB	$f22,t$	0	0	0	1
AABb	$f21,t$	0	0	$\frac{1}{2}$	$\frac{1}{2}$
AaBB	$f12,t$	0	$\frac{1}{2}$	0	$\frac{1}{2}$
AB/ab	$f11c,t$	$\frac{1}{2}(1-r)$	$\frac{1}{2}r$	$\frac{1}{2}r$	$\frac{1}{2}(1-r)$
Ab/aB	$f11R,t$	$\frac{1}{2}r$	$\frac{1}{2}(1-r)$	$\frac{1}{2}(1-r)$	$\frac{1}{2}r$

Then

$$\begin{aligned} g_{11,t+1} &= f_{22,t} + \frac{1}{2} f_{21,t} + \frac{1}{2} f_{12,t} + \frac{1}{2}(1-r) f_{11C,t} + \frac{1}{2} r f_{11R,t} \\ &= f_{22,t} + \frac{1}{2} f_{21,t} + \frac{1}{2} f_{12,t} - \frac{1}{2} r (f_{11C,t} - f_{11R,t}) \\ &= f_{22,t} + \frac{1}{2} f_{21,t} + \frac{1}{2} f_{12,t} + \frac{1}{2} f_{11C,t} - r d_t \end{aligned}$$

Where, according to equation d_t is defined as

$$d_t = \frac{1}{2}(f11c,t - f11R,t)$$

and

$$f22,t = f22,t-1 + \frac{1}{4} f21,t-1 + \frac{1}{4} f12,t-1 + \frac{1}{4}(1-r)^2 f11c,t-1 + \frac{1}{4} r^2 f11R,t-1$$

$$f21,t = \frac{1}{2} f21,t-1 + \frac{1}{2} r(1-r) f11c,t-1 + \frac{1}{2} r(1-r) f11R,t-1$$

$$f12,t = \frac{1}{2} f12,t-1 + \frac{1}{2} r(1-r) f11C,t-1 + \frac{1}{2} r(1-r) f11R,t-1$$

$$f11C,t = \frac{1}{2}(1-r)^2 f11C,t-1 + \frac{1}{2} r^2 f11R,t-1$$

$$f11R,t = \frac{1}{2} r^2 f11C,t-1 + \frac{1}{2}(1-r)^2 f11R,t-1$$

Thus

$$\begin{aligned} g11,t+1 &= f22,t-1 + \left(\frac{1}{4}+\frac{1}{4}\right) f21,t-1 + \left(\frac{1}{4}+\frac{1}{4}\right) f12,t-1 + [\frac{1}{4}(1-r)^2 \\ &+ \frac{1}{4} r(1-r) + \frac{1}{4} r(1-r) + \frac{1}{4}(1-r)^2] f11C,t-1 \\ &+ [\frac{1}{4} r^2 + \frac{1}{4} r(1-r) + \frac{1}{4} r(1-r) + \frac{1}{4} r^2] f11R,t-1^{-rdt} \\ &= f22,t-1 + \frac{1}{2} f21,t-1 + \frac{1}{2} f12,t-1 + \frac{1}{2}(1-r) f11C,t-1 \\ &+ \frac{1}{2} r f11R,t-1^{-rdt} \\ &= g11,t - r d_t \end{aligned}$$

(This equation is identical to Equation derived for the case of continued panmictic reproduction.)

$$g11,t+1 = g11,t - r_c d_t$$

Equation applies at continued self-fertilization. It is identical to Equation applying at continued panmictic reproduction. One should realize, however, that with panmictic reproduction the relation between $dt+1$ and dt was derived to be

$$d_{t+1} = (1-r_c)d_t$$

For autogamous reproduction, however, the relation between dt and $dt-1$ can be shown to be

$$d_{t+1} = \left(\frac{1-2r_c}{2}\right)d_t$$

Note: In the case of (continued) selfing, plants with a doubly heterozygous genotype, in the coupling phase or in the repulsion phase, can only be produced by doubly heterozygous parents, one can easily derive from Table

$$f_{11C,t+1} = 2\left(\frac{1-r}{2}\right)^2 f_{11c,t} + 2\left(\frac{r}{2}\right)^2 f_{11R,t}$$

that:

$$f_{11R,t+1} = 2\left(\frac{1-r}{2}\right)^2 f_{11R,t} + 2\left(\frac{r}{2}\right)^2 f_{11C,t}$$

Thus

$$f_{11,t+1} = \left[2\left(\frac{1-r}{2}\right)^2 + 2\left(\frac{r}{2}\right)^2\right](f_{11C,t} + f_{11R,t})$$

$$= \left(r^2 - r + \frac{1}{2}\right)f_{11,t} = \left[\left(r - \frac{1}{2}\right)^2 + \frac{1}{4}\right]f_{11,t}$$

Equations

$$d_{t+1} = \frac{1}{2}\left(f_{11,Ct+1} - f_{11,Rt+1}\right)$$

yield thus

$$d_{t+1} = \frac{1}{4}\left[(1-r)^2 - r^2\right]\left(f_{11,Ct+1} - f_{11,Rt+1}\right)$$

This gives Equation

$$d_{t+1} = \left(\frac{1-2r_c}{2}\right)d_t$$

$$d_t = \left(\frac{1-2r_c}{2}\right)^{t-1} d_t$$

Equations yield for the case of continued selfing:

$$g_{11,t+1} = g_{11,t} - r_c \left(\frac{1-2r_c}{2}\right) d_{t-1}$$

The parameter *dt* is still, as defined in Equation, equal to 1/ 2 (*f*11C,*t* – *f*11R,*t*). Equation shows that, unless *dt* = 0 or *rc* = 1/ 2, the haplotype frequencies will change from one generation to the next. The genotypic composition of F–, for F1 in coupling phase as well as in repulsion phase, depends directly on Equation, *viz.*

$$g_{11,\infty} = f_{22,\infty} = g_{11,1} - \left(\frac{2r}{1+2r}\right) d_1$$

Note: Combining the Equation yields in the case of continued selfing

$$g_{11,t+1} - g_{11,t} = -rd_1 \left(\frac{1-2r}{2}\right)^{t-1}$$

Repeated application of this equation result via

$$g_{11,2} - g_{11,1} = -rd_1 \left(\frac{1-2r}{2}\right)^{0}$$

$$g_{11,3} - g_{11,2} = -rd_1 \left(\frac{1-2r}{2}\right)^{1}$$

$$g_{11,t+1} - g_{11,t} = -rd_1 \left(\frac{1-2r}{2}\right)^{t-1}$$

$$g_{11,t+1} - g_{11,1} = -rd_1 \sum_{j=0}^{t-1} \left(\frac{1-2r}{2}\right)^{j}$$

The sum of the terms of this geometric series is

$$\frac{1-\left(\frac{1-2r}{2}\right)^{t-1}}{1-\left(\frac{1-2r}{2}\right)} = \frac{2}{1+2r}\left[1-\left(\frac{1-2r}{2}\right)^{t-1}\right]$$

Thus

$$g_{11,t+1} = g_{11,1-r}\left(\frac{2}{1+2r}\right).d_{1.}\left[1-\left(\frac{1-2r}{2}\right)^{t-1}\right]$$

Implying

$$g_{11,\infty}=f_{22,\infty}=g_{11,1}-\left(\frac{2r}{1+2r}\right)d_1$$

The quantity to be substituted in Equation for $d1$ amounts, according to Example, to 1/ 4 (1–2r) for F1 in the coupling phase and to –1/ 4 (1–2r) for F1 in the repulsion phase. Equation yields thus for F1 in the coupling phase:

$$g_{11,\infty}=f_{22,\infty}=\left(\frac{1-r}{2}\right)-\left(\frac{2r}{1+2r}\right)\left(\frac{1-2r}{4}\right)=\frac{1}{2(1+2r)}$$

For F_1 in the repulsion phase we get

$$g_{11,\infty}=f_{22,\infty}=\left(\frac{r}{2}\right)+\left(\frac{2r}{1+2r}\right)\left(\frac{1-2r}{4}\right)=\frac{2r}{2(1+2r)}$$

The genotypic composition of F– with regard to the complex genotype s for the two linked loci A-a and B-b (a) F_1 in coupling phase

	bb $\frac{2rc}{2(1+2r_c)}$	*Bb*	*BB*	
aa	0	0	$\frac{1}{2(1+2r_c)}$	$\frac{1}{2}$
Aa	$\frac{1}{2(1+2r_c)}$	0	0	0
AA	$\frac{1}{2}$	0	$\frac{2rc}{2(1+2r_c)}$	$\frac{1}{2}$
		0	$\frac{1}{2}$	1

(b) F_1 in repulsion phase

	bb $\frac{2rc}{2(1+2r_c)}$	*Bb*	*BB*	
aa	0	0	$\frac{1}{2(1+2r_c)}$	$\frac{1}{2}$
Aa	$\frac{1}{2(1+2r_c)}$	0	0	0
AA	$\frac{1}{2}$	0	$\frac{2rc}{2(1+2r_c)}$	$\frac{1}{2}$
		0	$\frac{1}{2}$	1

Table presents the genotypic composition of F–. It may be compared with Table presenting the genotypic composition obtained after continued panmixis.

In the case of linkage ($0 < rc < 1/2$) the frequencies of the haplotypes change in the course of the generations. For gametes with the *AB* haplotype the difference between $g11,1$ and $g11,-$ amounts to

$$g_{11,\infty} - g_{11,1} = \left(\frac{2r}{1+2r}\right)d_1$$

This amounts, according to Example, for F1 in the coupling phase to

$$\left(\frac{2r}{1+2r}\right)\left(\frac{1-2r}{4}\right) = \frac{r(1-2r)}{2(1+2r)}$$

And for F_1 in the repulsion phase to

$$\left(\frac{2r}{1+2r}\right)\left(\frac{2r-1}{4}\right) = \frac{r(2r-1)}{2(1+2r)}$$

These differences are for $0 < rc < 1\ 2$ generally quite small. For $rc = 1/4$, for instance, it amounts for F1 in the repulsion phase to $g11,1 - g11,- = 1/8 - 1\ 6 = -0.0417$.

We consider now the frequency of plants with a genotype obtained by crossing two parents. It may, for example, be desired to obtain genotype *AABB* from an initial cross of genotypes *AAbb* and *aaBB*.

The frequency of *AABB* plants amounts in population F2 to $f22,1 = 1/4\,rc/2$. Equation

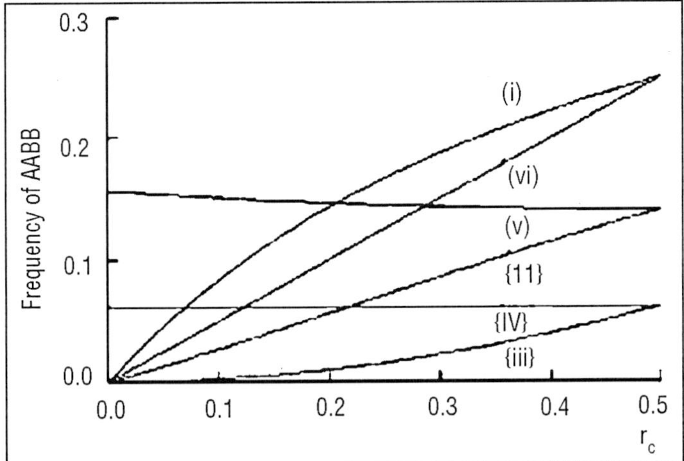

Fig. The Frequency of Plants with Genotype AABB as a Function of the Recombination Value rc.

Considered are population obtained by crossing of genotypes AAbb and aaBB followed by (i) continued self-fertilization until F-, (ii) selfing until F3, (iii) selfing until F_2, (iv) continued panmixis until linkage equilibrium, (v) continued panmixis followed by one round of reproduction by means of selfing, or (vi) doubling of the number of chromosomes in the gametes produced by F_1

yields for $t = 2$ the frequency of plants with genotype *AABB* in F3. When substituting the F2 genotype frequencies presented in Table one gets for an F1 in the repulsion phase:

$$f22,2 = \frac{1}{4}r^2 + \frac{1}{8}r(1-r) + \frac{1}{8}r(1-r) + \frac{1}{8}r^2(1-r)^2 + \frac{1}{8}(1-r)^2 r^2$$
$$= \frac{1}{4}r + \frac{1}{4}r^2 - \frac{1}{2}r^3 + \frac{1}{4}r^4$$

This amounts, for unlinked loci, to $f22,2 = 9\ 64 = _3/\ 8\ _2 = f00,2$. According to Equation the frequency of *AABB* plants in F– is $2r\ 2(1+2r)$. Because $2r\ 2(1+2r) \leq 1/\ 2(1+2r)$, plants with one of the parental genotypes will outnumber plants with this recombinant genotype to a greater extent as linkage is stronger, *i.e.* as *rc* is smaller.

In Figure curves (i), (ii) and (iii) show the values for *f*22 in F–, F3 and F2 as a function of *rc*. Recombination of alleles belonging to two different loci can only occur at meiosis of doubly heterozygous genotypes.

In populations of cross-fertilizing crops, doubly heterozygous genotypes tend to be permanently present; in populations of self-fertilizing crops they disappear.One should, however, be careful when speaking about 'the recombining effect of cross-fertilization'. This is illustrated for loci *A-a* and *B-b*. Continued panmictic reproduction gives eventually, at linkage equilibrium, $f22 = p2r2$. This amounts for $p = r = 1/\ 2$ to $1/\ 16$, whatever the recombination value. For tightly linked loci, with $rc < 1\ 14$, genotype *AABB* will indeed occur with a higher frequency in populations in linkage equilibrium than in populations obtained by continued selfing.

For less tightly linked loci, *i.e.* $rc > 1/\ 14$, the frequency of *AABB* will, however, be higher in *F*–. Thus one should not decide rashly to increase the frequency of plants with a recombinant genotype by the application of random mating in F2, F3,... populations of a self-fertilizing crop (Bos, 1977). With regard to unlinked loci continued random mating will only result in the genotypic composition of F2, because for unlinked loci the *F*2 population obtained by selfing will have the linkage equilibrium composition.

Selection in a cross-fertilizing crop is more efficient when increasing the frequency of homozygous recombinant genotypes by selfing. According to Note a single round of reproduction by means of self-fertilization in a population in linkage equilibrium gives

$$f_{22} = \frac{5 - 2r + 2r^2}{32}$$

Note: Consider a population in linkage equilibrium. It is obtained by panmictic reproduction starting with a single-cross hybrid variety. With regard to loci *A-a* and *B-B* a single round of reproduction by means of selfing results, according to Equation, in the following frequency of plants with genotype *AABB*:

$$f_{22} = \frac{1}{16} + \frac{1}{4}.\frac{1}{8} + \frac{1}{4}.\frac{1}{8} + \frac{1}{4}r^2.\frac{1}{8} + \frac{1}{4}(1-r)^2.\frac{1}{8} = \frac{5-2r+2r^2}{32}$$

For r = 1/ 2 this amounts to 9 64, *i.e.* _3/ 8 _2. It is the same value as obtained, from Equation, for an F3. The single reproduction by means of selfing gives thus the genotypic composition of an F3. This illustrates that the genotypic composition of the population in linkage equilibrium is equal to the genotypic composition for pairs of unlinked loci in an F2.

In a diploid crop, doubling the number of chromosomes of haploid plants is the fastest way to attain complete homozygosity. The frequency of plants with the desired recombinant genotype then amounts to 1/ 2 *rc*, *i.e.* 2/ *rc* times as high as in F2.

The frequency of doubly heterozygous plants is greatly reduced with reproduction by means of selfing. Depending on the recombination value, a single round of selfing reduces this frequency to only 1/ 4 to 1/ 2 of the frequency of plants with the *AaBb* genotype in the preceding generation. Note shows that the remaining portion of doubly heterozygous plants amounts to *f*11,*t*+1 *f*11,*t* = (*r* – 1/ 2)2 + 1 4, which amounts to 1/ 4 for *rc* = 1/ 2 and to 1/ 2 for *rc* = 0.

This reduction of the frequency of heterozygous plants is even stronger for more complex genotypes: a single round of selfing reduces the frequency of the complex genotype consisting of a heterozygous single-locus genotype for each of *k* unlinked loci to the portion (1/ 2)*k* of its preceding value.

CONCLUSION

Inbreeding occurs if mating plants are, on the average, *more* related than random pairs of plants. A more than average relatedness of the mating plants is thus a prerequisite. Relatedness implies, of course, that the plants involved share one or more ancestors. The strength of the inbreeding depends on the degree of relatedness of the mating plants. It has already been noted that mating of related plants may occur in random mating, but in that case it occurs as a matter of chance.

Note: Several yardsticks for measuring the degree of relatedness exist, a common one being the probability that an allele of a certain locus in some plant is identical by descent to an arbitrary allele at that same locus in its mate. In regular systems of inbreeding the degree of relatedness of the mating plants is uniform across all pairs of mating plants. In this book no attention is given to the determination of the degree of relatedness.

Regular systems of inbreeding are far more common in plant breeding than irregular systems. No attention will, therefore, be given to irregular systems of inbreeding. The counterpart of inbreeding is outbreeding. With outbreeding mating plants are on the average *less* related than random pairs of plants. Selfincompatibility is a natural cause for outbreeding as related plants tend to have a similar genotype at the incompatibility locus/loci. After intercrossing, such plants will produce no (or few) offspring. Artificial forms of outbreeding are

- Bulk crossing of two unrelated populations
- Selection of parents to be crossed in such a way that inbreeding is avoided as much as possible Outbreeding occurs also in the case of immigration.

The population genetic effect of inbreeding is a decrease in the frequency of heterozygous plants. This involves all loci, for all traits. (Random mating, on the other hand, is a mode of reproduction that may occur for certain traits and may simultaneously be absent for other traits). When starting with an F2 population and considering segregating loci, the frequency of heterozygous plants is the same for all loci. This applies to the successive generations of the superpopulation. Each subpopulation consists of few plants: in the case of selfing only a single plant, in the case of full sib mating only pairs of plants. Within these separate subpopulations reproduction is by means of random mating.

The random variation of the gene frequencies occurring in small populations causes the subpopulations to vary with regard to the frequencies of heterozygous plants: not only for different loci, but also for the same locus. Individual plants of the F2 (or F3, etc.) populations vary therefore in the number of heterozygous loci.

In diploid crops procedures for the production of doubled haploid lines (DH-lines) allow the production of pure lines from heterozygous parents in a single generation.

Doubling of the number of chromosomes of haploid plants, generated by parthenogenesis or by anther culture, yields immediately complete homozygosity. For dioecious crops as well as for self-fertilizing crops with a long juvenile phase, *e.g. Coffea arabica* L., this approach is an attractive alternative to continued inbreeding.

Tissue culture techniques for the regeneration of plants from anthers or microspores have been developed, for example in wheat, barley, rice and oilseed rape. Also elimination of paternal chromosomes, occurring when making *Hordeum vulgare* L. × *H. bulbosum* L. or *Triticum aestivum* L. × *Zea mays* L. crosses, permits production of DH-lines. (The paternal chromosomes are lost in a few cell divisions of the hybrid zygote/embryo.)

Note. DH-lines are mostly obtained directly from the gametes produced by the F1-plants. This has a few drawbacks

- Recombination is restricted to the F1 meiosis
- The proportion of DH-lines that are rejected because of poor performance is high. This is undesirable because of the cost of producing DH-lines.

To avoid these drawbacks one may use gametes from plants obtained by backcrossing the F1 or one may use F2- or even F3-plants. (The latter allows selection among F2-plants, followed by selection among F3-lines in the seedling stage). *In vitro* selection among the haploid embryos appeared to be feasible: the size and degree of embryo differentiation predicted which embryos would produce vigorous seedlings. Additionally the growth rate of the embryos was positively correlated with yield performance in the field $r = 0.3$, but this has found little practical application).

Continued self-fertilization is the natural mode of reproduction of selffertilizing crops. There are many economically important self-fertilizing crops. A number of these are

Barley *Hordeum vulgare* L.
Oats *Avena sativa* L.
Wheat *Triticum aestivum* L.
Rice *Oryza sativa* L.
Sorghum *Sorghum bicolor* (L.) Moench.
Finger millet *Eleusine coracana* (L.) Gaertn.
Pea *Pisum sativum* L.
Cowpea *Vigna unguiculata* (L.) Walp.
Dry bean *Phaseolus vulgaris* L.
Soybean *Glycine max* (L.) Merr.
Peanut *Arachis hypogaea* L.
Cotton *Gossypium* spp.
Arabica coffee *Coffea arabica* L.
Lettuce *Lactuca sativa* L.
Tomato *Lycopersicon esculentum* Mill.
Okra *Abelmoschus esculentus* (L.) Moench.
Sweet pepper *Capsicum annuum* L.

Self-fertilization is not always 100per cent in most of these autogamous crops, *e.g.* cotton, okra, sorghum. (The amount of outcrossing in sorghum is about 6per cent.) consider the genotypic composition of populations reproducing by a mixture of self-fertilization and cross-fertilization. Breeders regularly apply inbreeding in cross-fertilizing crops. They may have various reasons for doing this:

- The development of pure lines (mostly by continued selfing) for use as parents in the breeding of hybrid varieties, *e.g.* in maize or cucumber.
- To promote the efficiency of elimination of an undesired recessive gene
- Maintenance of a genic male sterile 'line'

Note:FS-mating occurs also when a maintaining a genic male sterile barley 'line': male sterile plants are harvested after having been pollinated by their male fertile full sibs.

(This is also applied in the case of recurrent selection in self-fertilizing cereals). Thus the harvesting of a female plant (say genotype *mm*) implies harvest of seed due to the cross *mm* ×*Mm* (where *Mm* represents the genotype assumed for hermaphroditic plants).

The genotypic composition of the obtained FS-family is (1/ 2, 1/ 2, 0). Repeated application of this procedure implies repeated FSmating. The most powerful form of inbreeding of cross-fertilizing crops, *e.g.* dioecious crops, occurs with repeated crossing of the type

- Full sib × full sib, *i.e.* full sib mating, or
- Parent ×× offspring.

Full Sib Mating

The offspring due to a cross of two genotypes constitutes a family. The plants belonging to the family share both their maternal and their paternal parent.

With regard to each other these plants are full sibs. Together they form a full sib family (FS-family).

Crossing of plants belonging to the same FS-family is called full sib mating (FS-mating). FS-mating may be used when inbreeding of dioecious crops, such as spinach or asparagus, is the aim.

It occurs spontaneously in the case of open pollination within FS-families grown in isolation. This is applied in hermaphroditic, monoecious or dioecious crops in the case of separated FS-family selection. Note describes how FS-mating is applied when maintaining a genic male sterile 'line'.

Parent ×× offspring mating

In this book the notation A ×× B indicates the cross A×B and/or the reciprocal cross B × A. Parent ×× offspring crosses, *i.e.* so-called PO-mating, can only be applied to perennial crops such as oil palm (producing gametes from the age of 4–5 years for many years) or asparagus (with a juvenile phase lasting two years).

The parent is still alive when its offspring reach the reproductive phase.

Repeated backcrossing implies continued application of crosses of the type 'recurrent parent ×× offspring'.

In the absence of selection the genotype of the offspring becomes identical to the genotype of the recurrent parent (if the recurrent parent has a homozygous genotype) or to the genotypic composition of the possible lines obtained by selfing of the recurrent parent (if the recurrent parent is heterozygous).

In this chapter only loci segregating for not more than two alleles per locus will be considered. For an extensive treatment of the population genetics theory of inbreeding the reader is referred to Allard, Jain and Workman (1968).

5

Evolution of Disassortative and Assortative Mating

REPEATED BACKCROSSING

A breeder may wish to improve an otherwise acceptable genotype by the incorporation of a specific major gene. For example

- It may be desired to improve the resistance of a rice variety or a lettuce variety against a new race of some disease.
- When breeding a hybrid variety it might be useful to develop a male sterile pure line which is genotypically identical to the pure line used as the paternal parent of the hybrid, except for its idiotype at the locus and cytoplasm controlling pollen development. Then one should transform the male fertile pure line into a male sterile line. This is done by pollination of a male sterile line by the paternal pure line parent. The obtained progeny is repeatedly, *i.e.* generation after generation, backcrossed with the male fertile pure line. (The latter line is called: maintainer line. It is, of course, maintained by continued selfing. In Note a somewhat different procedure for maintaining a male sterile line was mentioned, *viz.* full sib mating followed by harvesting of the male sterile plants. This procedure is applied with recurrent selection in self-fertilizing crops).

The genotype to be improved is called (for reasons that will become clear hereafter): recurrent parent. It may be a pure line (possibly a variety of a self-fertilizing crop or a pure line used in the production of a hybrid variety of a cross-fertilizing crop) or a clone. The allele determining the desired trait is designated by R.

It belongs to locus R-r and is to be incorporated into the recurrent parent. The latter is therefore crossed with a donor line containing the desired allele, but otherwise resembling the recurrent parent as much as possible. For all loci for which the recurrent parent and the donor line have a different genotype (save locus R-r), one wants to retain the genotype of the recurrent parent.

These loci may or may not be linked with locus *R-r*. With the introduction of the desired allele *R*, alleles belonging to other loci – which are possibly linked to locus *R-r* – are introduced as well. This phenomenon is called linkage drag. Many of these unintentionally introduced alleles will be undesirable. Often the breeder is not even aware of the introduction of such undesirable alleles, *e.g.* alleles belonging to loci controlling bitterness of the seeds).

Repeated backcrossing of the material under development with the recurrent parent, is applied in order to replace the dragged alleles step by step with the alleles of the recurrent parent. In this way a so-called near isogenic line is developed.

The rate of the replacement is considered for the simple situation of dominance of the desired allele, to be introduced from the donor, over the recurrent parent allele that is to be replaced. Each of all the other loci, for which a possibly unfavourable allele was introduced, is represented by locus *B*-β.

The actual (and favoured) genotype of the variety is represented by *BB*; the genotype of the donor by ββ. For the time being it is assumed that selection is only applied with regard to the trait controlled by locus *R-r*. Then it does not matter which allele of locus *B*-β is dominant, or whether the locus controls a trait that is expressed before or after pollen distribution. The recombination value for loci *R-r* and *B*-β is *rc*. Its value depends on the specific locus which is represented by *B*-β. For most loci *rc* will amount to 1/ 2. The slower the (rate of) replacement of allele β by allele *B*, the higher the number of backcross generations required to restore genotype *BB* for all loci represented by *B*-β.

Allele *R* is introduced by crossing the recurrent parent (say P1, with genotype *rB*/*rB*) with a donor (say P2, with genotype *R*β/*R*β). The obtained F1 has genotype *rB*/*R*β. The haplotypic composition of the gametes produced by F1 is

Haplotype

$$\begin{array}{lcccc} & rB & r\beta & RB & R\beta \\ f & \frac{1}{2}(1-r_c) & \frac{1}{2}r_c & \frac{1}{2}r_c & \frac{1}{2}(1-r_c) \end{array}$$

The first backcross, $P_1 \times F_1$*, result in a population (usually designated as* BC_1*) with genotypic Composition:*

Genotype

$$\begin{array}{lcccc} & rB/rB & r\beta/rB & RB/rB & R\beta/rB \\ f & \frac{1}{2}(1-r_c) & \frac{1}{2}r_c & \frac{1}{2}r_c & \frac{1}{2}(1-r_c) \end{array}$$

Elimination of plants with genotype *rr* transforms population BC1 into population BC1_. The genotypic composition of BC1 and the haplotypic composition of the gametes produced by each genotype in BC1are

Genotypic composition of BC_1

Genotype	*f*
$RB/r\beta$	r_c
$R\beta/rB$	$1-r_c$

Haplotypic composition of the gametes produced by each genotype

rB	$r\beta$	RB	$R\beta$
$\frac{1}{2}$	0	$\frac{1}{2}$	0
$\frac{1}{2}(1-r_c)$	$\frac{1}{2}r_c$	$\frac{1}{2}r_c$	$\frac{1}{2}(1-r_c)$

Haplotypic composition of the gametes produced by BC_1'as a whole is

Haplotype

	rB	$r\beta$	RB	$R\beta$
f	$\frac{1}{2}r_c+\frac{1}{2}(1-r_c)^2$	$\frac{1}{2}r_c(1-r_c)$	$\frac{1}{2}r_c+\frac{1}{2}r_c(1-r_c)$	$\frac{1}{2}(1-r_c)^2$

The second backcross, *i.e.* P1×× BC1, yields population BC2 with genotypic composition:

Genotype

	rB/rB	$r\beta/rB$	RB/rB	$R\beta/rB$
f	$\frac{1}{2}r_c+\frac{1}{2}(1-r_c)^2$	$\frac{1}{2}r_c(1-r_c)$	$\frac{1}{2}r_c+\frac{1}{2}r_c(1-r_c)^2$	$\frac{1}{2}(1-r_c)^2$

Because all BC1-plants have genotype *Rr*, half of the BC2-plants will have genotype *rr*. Elimination of the latter plants yields population BC2 with genotypic composition:

Genotype

	RB/rB	$R\beta/rB$
f	$1-(1-r_c)^2$	$(1-r_c)^2$

Likewise, population BC_*t* contains genotype *R*β/*rB* with frequency (1–*rc*)*t*. The frequency of plants with genotype *R*β/*rB* in population BC*t*_ is thus (1 – *rc*)*t*. For *rc* = 1/ 2 this amounts to (1/ 2)*t*. The frequency of genotype *RB*/*rB* amounts then to 1 – (1/ 2)*t*. The probability that a line, obtained by selfing in population BC*t*_ a random plant, might segregate for locus *B* –β is (1–*rc*)*t*.

We consider now the K unlinked loci $B1 - \beta1, B2 - \beta2, ..., BK - \beta K$. Locus R-r is not linked with any of these. Then in population BC_t the frequency of plants with the desired complex genotype will amount to

$$\left[1-\left(\frac{1}{2}\right)^t\right]^K = \left[\frac{2^t-1}{2^t}\right]^K$$

This expression is equal to Expression, tabulated in Table for $K = 1, ..., 14$ and $t = 1, ..., 7$. When considering $K = 7$ loci Table shows that in population BC5_ the frequency of plants with the complex genotype $RrB1B1B2B2 ... B7B7$ amounts to 0.801. In population BC6_ it is already 0.896. When considering $K = 14$ loci the frequency of plants with genotype $RrB1B1 ... B14B14$ amounts to 0.641 in population BC5_ and to 0.802 in population BC6_.

The frequency of plants with a complex genotype deviating for one or more of the loci $B1$-$\beta1$, ..., BK-βK from the genotype of the recurrent parent will amount to:

$$1-\left[\frac{2^t-1}{2^t}\right]^K$$

This equation gives the probability that a line, obtained by selfing a random plant taken from population BCt_, might segregate for one or more of the K loci.

Such segregation will also appear from a difference, for at least one trait, between plants of the line and the recurrent parent.

It may be concluded that, even for unlinked loci, five generations of backcrossing yield an insufficient reduction in the frequency of plants containing at one or more loci an undesired allele.

One or more additional backcross generations already implies a considerable reduction, especially for 'large' values for K. One should, of course, minimize K.

This can be done by using as the donor a genotype that resembles the recurrent parent as closely as possible.

An additional criterion for choosing a donor, follows from the dominance relationships among the alleles at the B-β loci. With regard to loci for which the recurrent parent allele B is not dominant over the donor allele β, one might distinguish, among the plants with genotype Rr, plants with genotype $RrBB$ and plants with genotype $RrB\beta$.

Selection of plants with genotype $RrBB$ implies then elimination of allele β.

Selection, particularly marker-assisted selection, among the plants with genotype Rr, of plants with the genotype of the recurrent parent (BB) reduces consequently the number of backcross generations required to attain the desired frequency of plants with genotype *RrBB*. Markers strongly linked to locus *B*-β and/or locus *Rr* are particularly useful. Among donor lines which differ from the recurrent parent with regard to their genotype for *K* loci, one should choose the donor containing a dominant allele at the highest number of these loci. Different donor lines can, in this respect, be compared by considering the similarity of the F1 and the donor: the greater the similarity, the larger the number of dominant donor alleles.

Until now the recurrent parent was assumed to have a homozygous genotype. When dealing with vegetatively propagated crops (such as apple, rhubarb, shallots, asparagus) the recurrent parent may be heterozygous for some locus *B*-*b*-β. The cross between the recurrent parent (with genotype *Bb*) and a donor (with genotype ββ) yields an F1 with the following genotypic composition

Genotype

	$B\beta$	$b\beta$
f	$\frac{1}{2}$	$\frac{1}{2}$

The frequencies of genotypes and alleles in BC_1', BC_2' and BC_3' then amount to:

	Genotype					Allele		
	bb	Bb	BB	$b\beta$	$B\beta$	b	B	β
f in BC_1' :	$\frac{1}{8}$	$\frac{1}{4}$	$\frac{1}{8}$	$\frac{1}{4}$	$\frac{1}{4}$	$\frac{3}{8}$	$\frac{3}{8}$	$\frac{1}{4}$
in BC_2' :	$\frac{3}{16}$	$\frac{3}{8}$	$\frac{3}{16}$	$\frac{1}{8}$	$\frac{1}{8}$	$\frac{7}{16}$	$\frac{7}{16}$	$\frac{1}{8}$
in BC_3' :	$\frac{7}{32}$	$\frac{7}{16}$	$\frac{7}{32}$	$\frac{1}{16}$	$\frac{1}{16}$	$\frac{15}{32}$	$\frac{15}{32}$	$\frac{1}{16}$

It will be clear that repeated backcrossing to a heterozygous recurrent parent is expected to result in a BC–' population with genotypic composition;

Genotype

	bb	Bb	BB
f	$\frac{1}{4}$	$\frac{1}{2}$	$\frac{1}{4}$

With regard to locus *B*-*b*-β. BC–_ is thus not identical to the recurrent parent, but to its S1 lines. The same applies to the two loci *B*1-*b*1-β1 and *B*2-*b*2-β2, which may be linked or not, if the genotype of the recurrent parent is *B*1*b*1*B*2*b*2.

Bos (1980) considered backcrossing in autotetraploid crops. In population BC*t*_ the frequency of plants containing the unintentionally introduced allele β was derived to be (1/ 2)*t*–1 if loci *R*-*r* and *B*-β are unlinked. Thus, compared with diploid crops, one additional backcross generation is required in order to obtain the same degree of replacement of β by *B*.

QUALITATIVE VARIATION

In the case of qualitative variation the relation between genotype and phenotype is more direct than in the case of quantitative variation: the classification of plants according to their phenotype tends to reflect the underlying genotypes. The population genetic effect of assortative mating resembles then the population genetic effect of selfing and the frequency of heterozygous plants decreases rather fast.

QUANTITATIVE VARIATION

With quantitative variation the relation between genotype and phenotype is disturbed by variation in the quality of the growing conditions: in that situation it is impossible to classify plants on the basis of their phenotype in such a way that all plants in some class have the same genotype, or to distinguish genotypes in such a way that all plants with a specified genotype belong to the same class of phenotypes.

In addition, the same phenotype can be produced by a wide range of different genotypes and thus, from both causes, it implies only a loose relationship between phenotype and genotype, which rules out attainment of complete homozygosity by means of continued assortative mating.

For both categories of variation the relation between genotype and phenotype is additionally disturbed by dominance, because different genotypes may then give rise to the same phenotype.

Disassortative mating implies crossing of plants belonging to different phenotypic classes; especially the two extreme classes. It may result in plant material with phenotypes mainly distributed around the mid-parent value. Maintenance of small populations, *e.g.* accessions in a gene bank, requires care to prevent inconspicuous change of the genotypic composition, due to random variation of the allele frequencies. Disassortative mating of early flowering plants with late flowering plants may be applied to maintain the typical average flowering time of some accession. In natural populations plants with extreme phenotypes, *e.g.* very early flowering plants and very late flowering plants, may have a reduced fitness.

Mating of plants with a different sex may be considered as disassortative mating.

Some authors classify the phenomenon of incompatibility among disassortative mating. Two forms of incompatibility may be distinguished: homomorphic and heteromorphic. In contrast to heteromorphic incompatibility, homomorphic incompatibility is not associated with anatomical differences. In cabbages homomorphic incompatibility is used to produce hybrid varieties. Heteromorphic incompatibility may occur as heterostyly, *e.g.* in primrose (*Primula* sp.). This provision indeed leads to disassortative mating with regard to flower structure.

Note: In primrose and buckwheat (*Fagopyrum esculentum* Moench.) heterostyly occurs: there are short-styled plants ('thrum') and long-styled plants ('pin'). Darwin noted that *Primula* spp. plants are pollinated by bees or moths possessing a long proboscis. If an insect collects nectar from a plants producing the *thrum* type of flowers it will pick up pollen around the base of its proboscis. Upon further feeding this pollen may be deposited on the long stigma of plants producing the *pin* type of flowers.

If so, the insect may pick up pollen near the tip of its proboscis. This might later be deposited on the short stigma of *thrum* flowers of other plants. The heterostyly is in fact associated with sporophytic self-incompatibility.

Primrose and buckwheat are thus both obligatory allogamous crops. Often two populations that compensate each other with regard to the expression for one or more traits are crossed. The aim of this initial cross is to introduce from one parent the gene(s) inducing a desired expression for some trait into the other parent, which is an otherwise acceptable genotype (or population).

The initial cross is followed by a programme of repeated backcrossing, in which plants with the improved expression are, generation after generation, selected to be crossed with the parent to be improved. Because of the disassortative mating involved in this procedure, repeated backcrossing is treated in this chapter. In fact disassortative mating is a mode of reproduction that may occur within some populations. Repeated backcrossing could therefore also have been considered, where bulk crossing was introduced.

In some crops sexual dimorphism occurs. It is possible that each plant can be classified as either a female or as male plant (this situation is called dioecy); or one may distinguish female plants and hermaphroditic plants, which may be monoecious or not.

IN THE END WE CAN SAY

Assortative mating occurs if intermating plants tend to resemble each other more, with regard to some trait, than two random plants. It implies a positive correlation between the mating plants of their phenotypic values for the trait involved. The genotypes of these plants for the loci controlling the expression for the trait will therefore tend, in general, to be similar. With disassortative mating, the mating plants will have a negative correlation of their phenotypic values for the considered trait: the mating plants tend to resemble each other less than random plants.

It is obvious that the trait involved in the resemblance should be expressed before pollen distribution. Thus assortative and disassortative mating are only conceivable for traits such as colour of hypocotyls (*e.g.* in radish, *Raphanus sativus* var. *radicula* L.), flower colour (*e.g.* in Brussels sprouts, *Brassica oleracea* L. var. *gemmifera* DC., Example), anther colour (*e.g.* in maize, *Zea mays* L.), number of tillers (*e.g.* in rye, *Secale cereale* L.), date of flowering.

Example: When producing hybrid seed of Brussels sprouts, by making use of sporophytic self-incompatibility, rows of plants representing inbred line A, with genotype *SaSa*, are intermixed with rows of plants representing inbred line B, with genotype *SbSb*. The pure lines involved may differ with regard to shape and size of the ultraviolet-coloured honey guide (which is invisible for the human eye). However, bees, responsible for the pollination, observe such differences. They tend to visit either flowers of the *SaSa* pure line or flowers of the *SbSb* pure line. Thus the bees apply assortative mating, which is counter-productive when the aim is to produce hybrid seed.

Example: Assortative mating occurs in cross-fertilizing crops, *e.g.* perennial ryegrass (*Lolium perenne* L.), spontaneously with regard to date of flowering. This phenomenon has attracted a lot of attention in ecological population genetics. The rare, very early flowering plants on the one hand, and the rare, very late flowering plants, on the other hand, are then at a disadvantage. In the case of self-incompatibility, these plants will have a reduced seed-set, due to the scarcity of nearby cross-compatible plants. Such selection against both extreme phenotypes is called stabilizing selection.

Plants may produce flowers over an extended period of time. This applies especially to wild plant species, but also to certain cultivated grass species or rye, certainly when grown at a low plant density. The crossing between flowers, or inflorescences, flowering at the same time does then, due to the overlap of flowering periods of different plants, imply rather imperfect assortative mating.

Some authors, *e.g.* Allard and Strickberger, have used the term 'phenotypic assortative mating' when considering the present form of assortative mating. They used the term 'genotypic assortative mating' where this book deals with inbreeding. It is questionable whether it is useful to distinguish between two forms of assortative mating: phenotypic resemblance implies at least some genotypic resemblance, especially in the case of qualitative variation. Li used the terms 'positive' and 'negative assortative mating' instead of assortative and disassortative mating.

The population genetic effect of assortative mating with regard to some trait is a decreased frequency of plants with a heterozygous genotype for the loci affecting the trait, as well as their linked neighbours. Experience shows that for loci controlling traits that have no relationship with fitness, a decreased frequency of plants with a heterozygous genotype is not associated with inbreeding depression. Inbreeding gives for all loci a decrease in the frequency of plants with a heterozygous genotype and so affects fitness traits and so may result in inbreeding depression. Assortative mating, however, exclusively decreases heterozygosity for loci controlling the expression for the trait involved in the resemblance.

Selection efficiency is promoted by an increased frequency of homozygous genotypes. Assortative mating may thus be a useful tool: in the case of self-incompatibility or dioecy a breeder could apply assortative mating to increase the frequency of homozygous plants, *e.g.* with respect to the locus controlling the colour of the hypocotyl of radish.

With qualitative variation the small number of different phenotypes can easily be distinguished. Thus for the colour of the hypocotyl of radish one may distinguish white and red. The plants can be classified according to the expression for the considered trait. The phenotypes of the plants belonging to the same class are equivalent.

Then, with assortative mating, the coefficient of correlation of the phenotypic values of the mating plants will approach the value 1. The rate of decrease of the frequency of plants heterozygous for the loci involved will then be similar to this rate in the case of self-fertilization. With quantitative variation the level of expression may behave as a continuous, random variable. This applies to traits such as single plant yield, plant height, or (to a lesser degree) date of flowering or number of tillers. Plants grouped into the same class of phenotypic values have roughly the same phenotype. In this case the coefficient of correlation of the phenotypic values of the mating plants will tend to be less than 1.

It should be clear that the rate of decrease of heterozygosity due to assortative mating strongly depends on the nature of the variation: qualitative or quantitative.

6

Genetic Resources for Food and Agriculture

EMBRYO TRANSFER

Embryo transfer (ET) involves the transfer of an embryo from a superior donor female to a less valuable female animal. A donor is induced to superovulate (produce several ova) through hormonal treatment. The ova obtained are then fertilized within the donor, the embryos develop and are then removed and implanted in recipient females for the remainder of the gestation period. Alternatively, the embryos can be frozen for later use.

HORMONAL TREATMENT IN AQUACULTURE

In the same way as female reproduction in livestock can be controlled by hormonal treatment, it is also an important tool in aquaculture where it is applied for 2 main purposes. The first is to control reproduction of fish and shellfish, primarily to induce the final phase of ova production in order to synchronize ovulation and to enable broodstock to produce fish early in the season or when environmental conditions suppress the spawning timing of females. Implants or injection of the hormonal compound are used extensively in salmon farming.

The second purpose is to develop monosex (single sex) populations, which can be desirable in many situations. This can be, inter alia, because one sex is superior in growth or has more desirable meat quality or to prevent sexual/territorial behaviour. For example, female sturgeon are more valuable than males because they produce caviar.

Female salmon are the more valuable sex, because sexually precocious males die before they can be harvested and salmon roe has an economic value. Male tilapia are more desirable than females because they grow twice as fast. In many fish and shellfish species, sex is not permanently defined genetically and thus it can be altered in a number of ways, including treatment with sexual hormones such as testosterone or estrogen derivatives in early stages of development. To develop all-male tilapia populations, methyltestosterone can be used while monosex trout can be produced using androgens.

SPERM/EMBRYO SEXING

In livestock, to get offspring of a desired sex (*e.g.* females are preferred for dairy animals, males for beef animals), separation of X and Y sperm (*e.g.* based on staining DNA with a fluorescent dye) for AI and sexing of embryos (*e.g.* using specific DNA probes) can be used. Although these technologies are being developed and refined in a number of research institutions, they are not used at the field level in any of the developing countries, except China.

CRYOPRESERVATION

Cryopreservation, referring to the preservation of germplasm in a dormant state by storage at ultra-low temperatures, usually in liquid nitrogen (-196 C), can be used to preserve biological material (*e.g.* seeds, sperm, embryos) of crop, livestock, forest or fish populations for potential use in the future.

The technology can be used for genetic improvement purposes and for management of genetic resources. In livestock, cryopreservation has been used in a number of developing countries for ex situ conservation of animal genetic resources, including Benin, Brazil, China, India and Kenya. In fish, cryopreservation of embryos is not possible but sperm cryopreservation works for many species and has been used in carp, salmon and trout breeding, especially when the aim is to "refresh" populations that have gone through a bottleneck.

Considering crops and forest trees, about 90per cent of the 6 million plant accessions in genebanks, mainly crops, are stored in seed genebanks. However, storage of seeds is not an option for crops or trees that do not produce seed, such as banana, or that produce recalcitrant or non-orthodox seed (*i.e.* seed that does not survive under cold storage and/or the drying conditions used in conventional ex situ conservation), such as mango, coffee, oak and several tropical forest tree species. In these situations, as well as for long-term storage of seeds from orthodox species, cryopreservation offers an alternative strategy for ex situ conservation, although its routine use is still limited.

Following plant cell, tissue or organ storage at low temperatures, plants can be regenerated. For various herbaceous (*i.e.* non-woody plants), hardwood (*i.e.* broadleaf, deciduous trees) and softwood species (*i.e.* coniferous trees), cryopreservation of a wide range of tissues and organs has been achieved. There is large scale application of shoot tip cryopreservation in fruit crop germplasm collections, such as in plum and apple. Seeds of most common agricultural and horticultural species can be cryopreserved.

TISSUE CULTURE-BASED TECHNIQUES

Tissue culture refers to the in vitro culture of plant cells, tissues or organs in a nutrient medium under sterile conditions. It has been widely used for over 50 years and is now employed to improve many of the most important developing country crops. There are a number of tissue culture-based technologies and they can be employed for a range of different purposes. Some of them, used with chromosome number manipulation.

MICROPROPAGATION

Micropropagation is the laboratory practice of rapidly multiplying stock plant material to produce a large number of progeny plants, using plant tissue culture methods. For instance, shoot tips of banana or potato are excised from healthy plants and cultivated on gelatinized nutrient media in sterile conditions (in test tubes, plastic flasks, or baby food jars), so that contamination with pests and pathogens is avoided. The obtained plantlets can be multiplied an unlimited number of times, by cutting them in single-node pieces and cultivating the cuttings in similar aseptic conditions. Millions of plantlets can be produced this way in a very short time.

The plantlets are then transplanted in the field or nurseries, where they grow and yield low-cost, disease-free propagation materials, ready to be distributed to farmers. Even if healthy plants are not available initially, specific in vitro techniques can also be applied to produce disease-free propagation material.

Today, micropropagation is widely used in a range of developing country subsistence crops including banana, cassava, potato and sweetpotato; commercial plantation crops, such as oil palm, coffee, cocoa, sugarcane and tea; niche crops such as cardamom and vanilla; and fruit trees such as almond, citrus, coconut, mango and pineapple. Some of the many countries with significant crop micropropagation programs include Argentina, Cuba, Gabon, India, Indonesia, Kenya, Nigeria, Philippines, South Africa, Uganda and Vietnam.

IN VITRO SLOW GROWTH STORAGE

Micropropagation procedures have been developed for over 1,000 plant species, many of which are today micropropagated commercially. The procedures include rapid multiplication, involving rapid growth and frequent subculture (regeneration) which is generally the objective of commercial micropropagation. Instead, the basis of successful in vitro storage of stock cultures is to increase the interval between subcultures by retarding the growth without any deleterious effects on the plants in culture. The strategy is used to conserve plant genetic resources and in vitro slow growth procedures can be used so that plant material can be held 1-15 years under tissue culture conditions with periodic subculturing, depending on the species. Normally, growth is limited using low temperatures often in combination with low light intensity or even darkness. Temperatures in the range of 0-5 °C are employed for cold-tolerant species and 15-20 °C for tropical species. Growth can also be limited by modifying the culture medium and reducing oxygen levels available to the cultures.

IN VITRO EMBRYO RESCUE

Wide crossing has only become possible by advances in plant tissue culture. A particular challenge was to overcome the biological mechanisms that normally prevent inter-specific and inter-genus crosses, as a high proportion of wide-hybrid seeds either do not develop to maturity or do not contain a viable embryo. To avoid spontaneous abortion, embryos are removed from the ovule at the earliest possible stage and placed into culture in vitro. Mortality rates can be high, but enough embryos normally survive the rigours of removal, transfer, tissue culture, and regeneration to produce adult hybrid plants for testing and further crossing.

First-generation, wide-hybrid plants are rarely suitable for cultivation because they have only received half of their genes from the crop parent. From the other (non-crop) parent they have received, not only the small number of desirable genes, but also thousands of undesirable genes that must be removed by further manipulation. This is achieved by crossing the hybrid with the original crop plant, plus another round of embryo rescue, to grow up the new hybrids. This 'backcrossing' process is repeated for about six generations (sometimes more), until a plant is obtained that is almost identical to the original crop parent, except that it now contains a small number of desirable genes from the non-crop parent plant. Wide-crossing programs can take more than a decade to complete, although MAS and anther culture can be used to speed up the process. Embryo rescue has been used occasionally in forest tree species, but its application is likely to be limited to a small number of hybrids of interest, which are sufficiently close to produce a normal embryo but where embryo development in vivo is a limiting factor.

MUTAGENESIS

This involves the use of mutagenic agents, such as chemicals or radiation, to modify DNA and hence create novel phenotypes. Induced mutagenesis has been used in crop breeding programs in developing countries since the 1930s. It also includes somaclonal mutagenesis, involving changes in DNA induced during in vitro culture. Somaclonal variation is normally regarded as an undesirable by-product of the stresses imposed on a plant by subjecting it to tissue culture. However, provided they are carefully controlled, somaclonal changes in cultured plant cells can generate variation useful to crop breeders. In forestry, use of somaclonal variation has been a popular subject for research, particularly during the 1980s, but the technology is generally seen to offer little for the genetic improvement of most major industrial forest tree species.

Almost 3,000 new crop varieties have been developed and released by countries using mutation-assisted plant breeding strategies and an estimated 100 countries currently use induced mutation technology. Case studies from Kenya (wheat), Peru (barley), sub-Saharan Africa (cassava) and Vietnam (rice) are described in IAEA (2008).

In the livestock sector, mutagenesis has also been used in developing countries. The sterile insect technique (SIT) for control of insects (*e.g.* screwworm and tsetse flies) relies on the introduction of sterility in the females of the wild population. The sterility is produced following the mating of females with released males carrying, in their sperm, dominant lethal mutations that have been induced by ionizing radiation. This method is usually applied as part of an area-wide integrated pest management approach and has been applied in developing countries in the livestock sector as well as for the control of crop pests. An estimated 30 countries use SIT against insect pests, including Chile and Peru.

Mutagenesis is also extensively used to improve the quality of microorganisms and their enzymes or metabolites used in food processing. The process involves the production of mutants through the exposure of microbial strains to mutagenic chemicals or ultraviolet rays. Improved strains thus produced are selected on the basis of specific properties such as improved flavour-producing ability or resistance to bacterial viruses.

FERMENTATION

Fermentation is the process of bioconversion of organic substances by microorganisms and/or enzymes of microbial, plant or animal origin. During fermentation, various biochemical activities take place leading to the break down of complex substances into simple substances and resulting in the production of a diversity of metabolites including simpler forms of proteins, carbohydrates, fats, such as sugars, amino acids, lipids, as well as new compounds such as antimicrobial compounds (*e.g.* lysozyme, bactericins); organic acids (*e.g.* lactic acid, acetic acid, citric acid); texture-forming agents (*e.g.* xanthan gum); and flavours (esters and aldehydes). Apart from the various new products that are yielded during fermentation, the process is widely known for its preservative benefits. The new products that emerge following fermentation have been found to have potential for longer shelf lives, and they have characteristics quite different from the original substrates from which they are formed.

Fermentation is globally applied to preserve a wide range of raw agricultural materials (cereals, roots, tubers, fruit and vegetables, milk, meat and fish, etc.). Commercially produced fermented foods which are marketed globally include dairy products (cheese, yogurt, fermented milks), sausages and soy sauce. Fermentation of sugars is also central to production of bioethanol from agricultural feedstocks.

Certain microorganisms associated with fermented foods, in particular strains of the Lactobacillus species, are probiotic *i.e.* used as live microbial dietary supplements or food ingredients that have a beneficial effect on the host by influencing the composition and/or metabolic activity of the flora of the gastrointestinal tract. They can also be used as feed additives for monogastric and ruminant animals, and have been applied for this purpose in China, India and Indonesia. In developing countries, fermented foods are produced generally at the household and village level, using traditional processes that are uncontrolled and dependent on spontaneous 'chance' microorganisms from the environment. Modern fermentation processes employ the use of well constructed vessels (fermenters/bioreactors), with appropriate controlled mechanisms for temperatures, pH, nutrients levels, oxygen tensions among others and also use selected microorganisms and/or enzymes for their operations.

BIOFERTILISERS

Soils are dynamic living systems that contain a variety of microorganisms such as bacteria, fungi and algae. Maintaining a favourable population of useful microflora is important from a fertility standpoint. The most commonly exploited microorganisms are those that help in fixing atmospheric nitrogen for plant uptake or in solubilizing/mobilizing soil nutrients such as unavailable phosphorus into plant-available forms, in addition to secreting growth-promoting substances for enhancing crop yield. As a group, such microbes are called biofertilisers or microbial inoculants. They can be generally defined as preparations containing live or latent cells of efficient strains of nitrogen-fixing, phosphate-solubilizing or cellulolytic microorganisms used for application to seed or soil with the objective of increasing the numbers of such microorganisms and accelerating certain microbial processes to augment the availability of nutrients in a form that plants can assimilate readily. Biofertilisers have been used in a number of developing countries, such as Kenya and Thailand, often involving nitrogen-fixing Rhizobia bacteria.

BIOPESTICIDES

Living organisms that are harmful to plants and cause biotic stresses are collectively called pests, and they cause tremendous economic damage to plant production worldwide. Biopesticides are mass-produced, biologically based agents used for the control of plant pests. They can be living organisms, such as microorganisms, or naturally occurring substances, such as plant extracts or insect pheromones. Microorganisms used as biopesticides include bacteria, protozoa, fungi and viruses and they are used in a range of different crops.

For example, different biopesticides are available for controlling locusts. As an illustration, a biopesticide containing spores of the fungus Metarhizium anisopliae, was used to control a migratory locust infestation in an FAO project in 2007 in Timor-Leste. Surveys revealed that an area of about 20,000 hectares was infested with gregarious nymphs and that there was a serious threat to the rice crop. The target area was considered unsuitable for chemical spraying because of high density human settlement and many water courses, so the infestation was treated with the biopesticide, targeting flying swarms using a helicopter, spraying in a time period of over one month. Note, biopesticides generally have a slow action compared to conventional chemicals and, for that reason, the latter are preferred if crops are under immediate threat.

BIOMASS FOR ENERGY PRODUCTION

Approximately 40 percent of all primary energy production, or photosynthesis, occurs in the seas. In this process, oceanic plants (phytoplankton, seaweeds, seagrasses) take up carbon dioxide (CO_2) and, with light energy from the sun, convert it into organic carbon (primarily sugars) and oxygen. The oceans contain 50 times as much carbon dioxide as does the atmosphere, and it is estimated that primary production incorporates 35 gigatons (1 gigaton = 1 x 10*15* grams) of carbon into marine biomass annually. This abundant source of fuel for energy production has not been tapped commercially because it is not competitive with soybean meal and other easily harvested, traditional sources of biomass, and also because, regardless of the source, biomass is not competitive with other types of fuels.

The Federal Government should continue to support research on the use of biotechnology to enhance biomass production and utility. At least three general approaches are being explored.

First, the enzyme that captures CO_2 for photosynthesis — ribulose bisphosphate carboxylase\oxygenase or "RUBISCO" — is relatively inefficient, so supercomputers are being used to verify structural information, and the enzyme is being redesigned to optimize its function. Second, the chemical composition of biomass can be altered to make it more suitable for particular applications.

For example, marine microalgae are being genetically engineered to boost their lipid content, with the aim of providing a source of alternative fuels that is more economical than are conventional sources. Third, biotechnology is being used to convert biomass to ethanol and other alternative forms of energy and chemical feedstocks.

PLANT GENOME RESEARCH

The U.S. agricultural research community has mounted a considerable effort to understand the structure, function, and regulation of genes in a wide range of crop and forestry species. There may never be resources sufficient to sequence completely the genomes of all agriculturally important plants, but there is a high degree of similarity among species. Therefore, it is cost-effective to select one representative plant for concerted sequencing of its entire genome.

The model experimental plant *Arabidopsis thaliana* offers the best chance for achieving a completely sequenced plant genome in the foreseeable future.

Arabidopsis has the smallest known genome of any flowering plant, with

100 million base pairs and little repetition in its DNA. Detailed genetic and physical maps have been developed, and understanding of the plant's biology is growing rapidly. A complete understanding of the *Arabidopsis* genome would offer enormous potential for improving agriculture. In addition to studies on *Arabidopsis*, parallel studies will be required at some level with plants used to grow food and fibre.

Knowledge of a single gene can have broad impact. An example is a recent breakthrough in the identification and sequencing of a tomato gene that confers resistance to a bacterial disease. The sequencing of this gene has given scientists the first unambiguous glimpse of the resistance response of a plant to one of its pathogens. This discovery has opened the way to identification of disease resistance genes in other plants as well as created an avenue for investigating the molecular basis for disease resistance.

GENETICS OF RHIZOSPHERE MICROORGANISMS

Most farms grow only three or four crops, and some grow the same crops year after year in the same fields. These practices promote soil infestations by pathogens that damage or destroy roots.

For most crops, soilborne plant pathogens generally are uncontrolled, except by limited crop rotations and some host plant resistance.

Recent studies show that some soil microorganisms associated with the roots of crops carry and express genes for defence of roots against pathogens. Little is known about these beneficial microorganisms, which represent an enormous untapped genetic resource for improving crop production and efficiency of fertilizer use. Identification and genetic studies of these microorganisms could enable their direct deployment or use of their genes in the development of disease-resistant crops.

NEW USES FOR PLANT MATERIALS

Higher plants are very efficient sources of renewable organic materials. In particular, corn starch, which is both a useful raw material and an important human and animal food, is produced in large quantities at about the same cost per pound as is crude petroleum. Many genes involved in starch biosynthesis have been cloned recently and it should be possible to explore the development of new kinds of starches for use in Non-food applications. The potential use of starch as a co-polymer in biodegradable plastics is being explored.

Plant storage lipids (*i.e.*, oils) can be produced in large quantities, and they represent a chemically versatile form of biomass. Enhancement of oil yield and composition in plants such as palm and rapeseed depends on increased understanding of the factors that regulate lipid biosynthesis. Progress has been made in identifying genes and enzymes involved in the early steps of fatty acid and lipid biosynthesis. To date, the most promising approach to manipulating oil yield and composition involves modifying the vast variety of fatty acids found as lipid constituents in various wild species. This is technically possible only if the basic biochemistry is elucidated.

NOVEL PRODUCTS FROM PLANTS

Higher plants synthesize an enormous spectrum of chemical constituents. Many have known value as drugs, biomaterials, solvents, flavorings, fragrances, or coloring agents. Through development of transgenic plants, the purity of many plant-derived chemicals can be enhanced and the range of chemicals expanded. One example is the pending development of plants that produce hydroxylated fatty acids. These oils can be used in applications ranging from hydraulic fluids to nylon synthesis. Similarly, the recent development of plants that accumulate biodegradable thermoplastics illustrates the potential for producing completely new compounds in plants.

Many natural chemicals are important in plant defence mechanisms or in abiotic stress responses. Some chemicals confer important horticultural qualities to plants. Thorough studies have been conducted on the biosynthesis of certain classes of these compounds, such as flavonoids, cyanogenic glycosides, and certain alkaloids. Other classes of compounds, such as terpenoids, have received relatively little attention. Terpenoids are important for plant growth and can be used as essential oils, resin acids, and pigments. An efficient mechanism must be developed to obtain specific biochemical information about these classes of compounds, to complement the extensive ongoing efforts to identify them and elucidate their structures. Biotechnology offers promise for accelerating the characterization of these substances as well as facilitating their production for commercial use.

GROWTH AND DEVELOPMENT

Priority: Extend understanding of the biochemical and molecular basis of growth and development including structural biology of plants and animals.

Biotechnology offers opportunities to study a wide variety of genetically regulated growth and development processes. Alteration of these processes can help improve product quality and nutritional and economic benefits. The genetic makeup of plants and animals can be altered by either insertion of new, useful genes or removal of unwanted ones. Discoveries made through biotechnological approaches are changing the way plants and animals are grown, boosting their value to growers, processors, and consumers alike.

The many opportunities in plant research. The numerous potential benefits range from the development of specialty starch polymers for packaging and oils for use as industrial lubricants to modification of post-harvest traits such as ripening and senescence. The recent development of tomatoes with modified ripening characteristics may lead to extended post-harvest shelf life of fruits, vegetables, and flowers. Information about the biological basis of cold tolerance may result in the development of cold-hardy flowers and ornamentals now grown only in subtropical or tropical climates. The engineering of increased salt tolerance in plants may offset losses due to saline soils.

CONCERNS ABOUT FOOD PRODUCTION

Some concerns about the use of biotechnology for food production include possible allergic reactions to the transferred protein. For example, if a gene from Brazil nuts that produces an allergen were transferred to soybeans, an individual who is allergic to Brazil nuts might now also be allergic to soybeans. As a result, companies in the United States that develop genetically engineered foods must demonstrate to the U.S. Food and Drug Administration (FDA) that they did not transfer proteins that could result in food allergies. When, in fact, a company attempted to transfer a gene from Brazil nuts to soybeans, the company's tests revealed that they had transferred a gene for an allergen, and work on the project was halted. In 2000 a brand of taco shells was discovered to contain a variety of genetically engineered corn that had been approved by the FDA for use in animal feed, but not for human consumption. Although several anti-biotechnology groups used this situation as an example of potential allergenicity stemming from the use of biotechnology, in this case the protein produced by the genetically modified gene was not an allergen. This incident also demonstrated the difficulties in keeping track of a genetically modified food that looks identical to the unmodified food. Other concerns about the use of recombinant DNA technology include potential losses of biodiversity and negative impacts on other aspects of the environment.

SAFETY AND LABELLING

In the United States, the FDA has ruled that foods produced though biotechnology require the same approval process as all other food, and that there is no inherent health risk in the use of biotechnology to develop plant food products. Therefore, no label is required simply to identify foods as products of biotechnology. Manufacturers bear the burden of proof for the safety of the food. To assist them with this, the FDA developed a decision-tree approach that allows food processors to anticipate safety concerns and know when to consult the FDA for guidance. The decision tree focuses on toxicants that are characteristic of each species involved; the potential for transferring food allergens from one food source to another; the concentration and bioavailability of nutrients in the food; and the safety and nutritional value of newly introduced proteins.

BIOTECHNOLOGY AND GLOBAL HEALTH

The World Health Organization estimates that more than 8 million lives could be saved by 2010 by combating infectious diseases and malnutrition through developments in biotechnology. A study conducted by the Joint Centre for Bioethics at the University of Toronto identified biotechnologies with the greatest potential to improve global health, including the following:

- Hand-held devices to test for infectious diseases including HIV and malaria. Researchers in Latin America have already made breakthroughs with such devices in combating dengue fever.
- Genetically engineered vaccines that are cheaper, safer, and more effective in fighting HIV/AIDS, malaria, tuberculosis, cholera, hepatitis, and other ailments. Edible vaccines could be incorporated into potatoes and other foods.
- Drug delivery alternatives to needle injections, such as inhalable or powdered drugs.
- Genetically modified bacteria and plants to clean up contaminated air, water, and soil.
- Vaccines and microbicides to help prevent sexually transmitted diseases in women.
- Computerized tools to mine genetic data for indications of how to prevent and cure diseases.
- Genetically modified foods with greater nutritional value.

GENE SEQUENCING AND MAPPING

PRIORITY: Continue mapping and sequencing of animal/plant/microbial genomes to elucidate gene function and regulation, and to facilitate the discovery of new genes as a prelude to gene modification.

To locate desired genetic traits of organisms and effectively use molecular probes to identify DNA sequences, researchers need precise and relatively complete genetic maps. The combination of detailed maps and DNA sequences of genomes can provide a guide to plant and animal breeders for the development of new, efficient, and informed breeding strategies.

The Federally funded Human Genome Project is rapidly advancing gene mapping and sequencing technologies as well as data management techniques for analysing genome data from humans and other organisms. These advances have the potential to make important contributions to the genetic improvement of crops and animals. The agricultural research community is in an ideal position to apply the technical and scientific advances from the Human Genome Project to address the biology of organisms important to agriculture, aquaculture, and forestry.

PLANT FLOWERING

Scientists are gaining their first glimpse into the complex mechanisms of flowering. Most plants use environmental cues to initiate flowering so that individuals of the same species flower synchronously, thus maximizing the chances of successful outcrossing. Scientists have discovered molecules in plants called phytochromes that sense dark and light conditions and send signals that regulate plant activities such as sprouting, growth, and flowering.

There is still much to learn about how flowers develop. Investigations of genes controlling floral meristems — the tissue that forms flowers — promise to help explain the genetic mechanisms that manage the establishment and maintenance of this tissue. Ongoing research also is extending understanding of the physiological processes that take place during the transition from vegetative to reproductive development.

MECHANISMS OF PLANT FERTILIZATION

From apples to zucchini, most foods originate from the fertilized flowers of plants. In tomatoes, for example, several genes have been discovered that act within a flower's anthers, where pollen is produced. Promoters (regions of genes that act as on-off switches) also have been identified. Scientists now are pinpointing the overall molecular mechanisms that cue tomato pollen to fertilize the female part of a tomato flower.

Increased understanding of plant fertilization eventually could lead to incremental improvements in the yield and energy efficiency of crop production.

Some plants have genes that render their anthers infertile. This male sterility allows production of higher yielding hybrid seeds, whereby male-sterile plants are fertilized with pollen from specially selected sources. Genetically engineered systems for controlling male fertility are used to produce hybrid seed of important crops such as oilseed rape, corn, and rice. The technology also is applicable to tomato, lettuce, and a wide range of other crop plants.

These advances eliminate the need for costly, labour-intensive hand or mechanical removal of anthers in making hybrid crosses and will have a major impact on the billion-dollar hybrid seed industry. In addition, the knowledge gained by studying anther development and fertilization at the molecular and genetic levels will reveal other approaches that can be used to produce novel varieties of hybrid crop plants.

BIOLOGICAL BASIS OF PLANT FORM

Conventional breeding often is aimed at achieving certain visual traits. Once the biological basis for these traits is understood, biotechnology could provide the tools to change plant architecture precisely and optimize plant form. For example, changes in leaf form and/or number could maximize photosynthetic capacity and increase production, and changes in root architecture could facilitate and maximize water and mineral capture and uptake. Also, the tools of biotechnology could be exploited to design crops that thrive on soils contaminated with heavy metals, such as cadmium or mercury. Scientists are studying and beginning to redesign the genes responsible for the activity of phytochelatins, natural compounds in plants that bind these metals and detoxify them.

ANIMAL GROWTH AND DEVELOPMENT

Research also is needed to explore the genetic basis of animal growth and development. Technologies have been developed for gene cloning, gene transfer, *in vitro* culturing, and sex determination of embryos, and these approaches are being refined for use with animals. Transgenic embryos or offspring have been produced from rabbits, chicken, fish, sheep, swine, and cattle. Genes can be targeted for expression in specific tissues, but the efficiency of gene transfer methods should be improved, and genes important for growth and development must be characterized further.

ENVIRONMENTAL INTERACTIONS

Priority: Elucidate the molecular basis of interactions of plants and animals with their physical and biological environments, as a basis for improving the organisms' health and wellbeing. Future gains in agricultural productivity and sustainability depend heavily on the use of biotechnology to improve the health and wellbeing of agriculturally important plants and animals. Research on environmental interactions is emerging rapidly as a powerful means of increasing resistance to stress and disease, with consequent economic and environmental benefits. For example, basic research on the genetic control of biochemical responses to water stress in drought-hardy native plants can suggest how crop plants might be modified genetically for more consistent performance in years of drought. Similar studies are needed to improve the tolerance of U.S. crops to heat stress, winter and frost damage, harm caused by salinity, and other physical stresses.

New crop genotypes also must be resistant to an expanding diversity of pests and diseases. Basic research at the organismal and cellular levels, together with classical breeding techniques, already has paid significant dividends to farmers, consumers, and the environment. Today, U.S. wheat, corn, soybean, barley, and sorghum are grown without fungicides because resistant varieties have been bred. However, these and most other U.S. crops still are subject to major damage from insects and diseases caused by soilborne plant pathogens and insect-vectored viruses, for which there has been no useful source of resistance. As a result, U.S. agriculture continues to depend on pesticides. The survival of high-value horticultural crops depends largely on soil fumigants.

Biotechnology also can help reduce the billions of dollars in agricultural losses caused by animal diseases. There are opportunities to boost the immune competency of animal hosts by genetic manipulation, develop biotechnology-derived regulators of immune function, and identify ways to prevent attachment of disease agents to host cells. Factors affecting the disease-producing capacity of a bacterium or virus can be identified, and strategies can be devised to help the host withstand these pathogens. Through the use of techniques such as enzyme-linked immunosorbent assays, polymerase chain reaction (PCR), and monoclonal antibody-based systems, biotechnology can contribute to major advances in diagnostics.

PLANT PESTS AND DISEASES

Virologists have transferred virus genes — such as those for production of virus coat proteins — to plants, thereby conferring resistance to those viruses in otherwise susceptible hosts. Some genes also have been shown to limit virus replication and associated damage.

Replication depends on the regulatory and metabolic processes of the host; these processes must be understood before scientists can engineer virus-resistant plant varieties. Another possible way to increase plant resistance to viral infections would be to interfere with the spread of the virus, but this approach would require an understanding of the biochemical mechanisms of spread and transmission. All crop plants are susceptible to bacterial diseases, which have been difficult to control through plant breeding. Bactericides are not a complete solution, either, because bacteria quickly evolve resistance to them. The bacterial disease fire blight is destructive to pears, apples, quince, and some ornamental plants. This disease has plagued fruit producers for over three centuries, but its mechanism of attack only recently has been revealed using the tools of molecular biology. This knowledge is expected to provide ways to combat this disease. Nearly all agriculturally important fruit, vegetable, and grain varieties acquire fungal diseases that can cost growers billions of dollars annually. Information is accumulating rapidly on the molecular basis of plant responses to infection by these pathogens. In many cases, the difference between resistance and susceptibility is the rate of response; if rapid, then the plant is resistant. Scientists have found that deliberately disarmed fungal pathogens can trigger the necessary response in susceptible (*i.e.*, slow responding) plants before the virulent pathogen arrives. This approach has opened a new avenue in disease control: the use of "pathogen derived" biocontrol agents to induce resistance in otherwise susceptible plants. This technique is only beginning to emerge as a way to control heretofore unmanageable plant diseases.

DNA PROBES

DNA probe kits have been developed for *Salmonella*, *Listeria*, *E.coli* 0157:H7, and *Staphylococcus aureus*. Compared to traditional culture-plating methods, these new diagnostic kits offer greater precision, shorter turnaround times, and reduced need for highly trained personnel.

DNA diagnostic techniques for Norwalk viruses also are commercially available. Culture techniques have not been successful in detecting these leading causes of gastroenteritis, making the availability of genetically based techniques critical to the detection and identification of the Norwalk viruses.

In addition to detecting food contaminants, DNA probes and other tools of biotechnology can help reduce levels of naturally occurring toxicants in foods. DNA probes can be exploited in research and plant breeding to isolate genes associated with the biosynthesis of major toxicants, facilitate understanding of the genetic regulatory mechanisms for toxicants in plants, and develop lines of plants with reduced levels of toxicants.

Mycotoxins in food are a periodic threat to food safety. DNA probes could help detect the presence of mycotoxin-producing fungi that grow under certain conditions in plant materials such as improperly dried corn and peanuts. DNA probes could also be used to learn more about the sources of fungal contamination in the environment and as a means to develop management strategies under field conditions.

BIOSENSORS

The development of biosensors offers great promise for improving food processing, analysis, and safety assurance. The highly specific actions of biological molecules can be exploited for use in biosensors that can measure the concentration of specific components in complex mixtures. Enzymes, antibodies, and microbial cells can be immobilized on solid surfaces, and the specific reactions they mediate can be detected by various physical and chemical means. Biosensors are commercially available to detect a variety of sugars, alcohols, esters, peptides, amino acids, cell types, and antibiotics. Development of tailor-made membranes capable of separating molecules based on size, electrical charge, or solubility will accelerate biosensor development as well as the exploitation of biomimetic systems.

Miniaturization and mass production of biosensors could increase their availability and decrease their unit cost. Technologies such as microlithography, ultrathin membranes, and molecular self-assembly have the potential to facilitate the miniaturization of molecular and cellular processes for enrichment, detection, and analysis of chemical and microbiological contaminants in food. Advances in the semiconductor industry have made it possible to combine chemical and biological components and integrated circuits in miniaturized systems. Biosensors can be inserted directly into food processing streams to obtain on-line, real-time measurements of important food processing parameters. Miniature biosensors also could be incorporated into food packages to monitor temperature stress, microbial contamination, or remaining shelf life, and to provide a visual indicator to consumers of product state at the time of purchase.

SYSTEMATIC STUDY OF FOOD-BORNE MICROBES

An appropriately designed long-term molecular biology study could characterize and correlate food-associated microbiological isolates in existing culture collections. Food-related isolates from these collections could be used to screen foods for microbiological contamination. Foods identified in this way then would serve as focal points for regulatory agencies and critical systems aimed at improving control over growing, shipping, processing, distribution, and other steps during which these foods may become contaminated.

AQUACULTURE

Priority: Use the tools of modern biotechnology to improve the health, reproduction, development, growth, and overall wellbeing of cultivated aquatic organisms; and promote the interdisciplinary development of environmentally sensitive, sustainable systems that will enable significant commercialization of aquaculture.

Aquaculture, which long has been practiced in Asia and is increasingly popular in the United States, Europe, and South America, will benefit tremendously from the use of new molecular tools and processes. With worldwide seafood demand projected to increase 70 percent in the next 35 years, and harvests from capture fisheries stable or declining, aquaculture will have to produce seven times as much seafood as it generates now to supply global demand by 2025.

The use of modern biotechnology to intervene in the rearing process and enhance production of aquatic species holds great potential not only to meet this demand, but also to improve U.S. competitiveness in aquaculture. The U.S. aquaculture industry has grown rapidly in recent years. Farm gate receipts exceeded $800 million in 1992 — a fourfold increase since the early 1980s.

The growth and international competitiveness of the U.S. aquaculture industry will be determined by the size of the resource investment in research and technology development. This investment should be made through a partnership of Federal and state agencies and the private sector. The Federal role is to provide leadership in supporting research to advance knowledge in important research areas and to facilitate the transfer of promising results and technologies to the private sector.

The major research issues in aquaculture are similar to those for other agricultural sectors, but the knowledge base for aquaculture is comparatively Meagre. Development of this knowledge is a particular challenge due to the diversity of cultured aquatic species and the systems for their production. Federal support for biotechnology research in this area will expand the knowledge base and yield significant dividends.

The application of biotechnology promises significant benefits to both producers and consumers of aquacultural products. The use of genetically enhanced organisms may improve production efficiency through improvements in growth rates, food conversion, disease resistance, and product quality and composition. The application of biotechnology to aquaculture also may help conserve wild species and genetic resources and provide unique models for biomedical research.

ENHANCING REPRODUCTION AND EARLY DEVELOPMENT

Biotechnology can be applied to enhance reproduction and early development of cultivated aquatic organisms. The resulting benefits could include year-round production of gametes and fry of economically valuable species and creation of new markets for specialized, genetically improved broodstock. Similarly, biotechnology may provide techniques for improving the reproductive success and survival of endangered species, thereby helping to preserve the diversity of life on Earth. As a first step, research should be directed towards improving basic understanding of environmental, hormonal, biochemical, and genetic control of reproduction.

More specifically, scientists must identify and understand the mechanisms of expression of genes involved in reproduction and development, improve technologies for cryo-preserving gametes and embryos, improve delivery systems for administration of natural and synthetic hormones, and enhance understanding of the pharmacokinetics of uptake and release of administered hormones.

IMPROVING HEALTH AND WELL-BEING

Biotechnology offers substantial opportunities to improve the health and wellbeing of cultivated aquatic organisms. More than 50 diseases affect fish and shellfish cultured in the United States, causing losses of tens of millions of dollars annually. Biotechnology not only can improve the survival, growth, vigour, and wellbeing of cultivated stocks, but also can reduce disease transfer between cultivated and wild stocks. New products and market opportunities can be developed related to aquatic animal health and wellbeing.

The tools of molecular biology can provide a basic understanding of host immunity, resistance, and susceptibility to diseases and associated pathogens by furnishing information about life cycles and mechanisms of pathogenesis, antibiotic resistance, and disease transmission. Improved technologies must be developed for detecting and diagnosing pathogens and diseases and for enhancing the genetic basis of disease resistance, thereby reducing the need for antibiotics and other drugs.

Potential products resulting from this research include gene therapy techniques; broodstock free of pathogens; safe, effective prophylactic agents, including immune modulators, antigens, and vaccines; safe, effective therapeutic agents; and improved systems for administering prophylactic and therapeutic agents.

IMPROVING QUALITY AND VALUE

The Federal Government has a responsibility to help ensure the safety and quality of food supplies, and biotechnology can and should be an invaluable tool in carrying out this mandate. Biotechnology can be employed to assess and improve the safety, freshness, colour, flavour, texture, taste, nutritional characteristics, and shelf life of aquacultural food products. In addition, practical technologies can be developed to detect and assay toxins, contaminants, and residues in seafood, and to reduce or eliminate contaminants.

There are also opportunities to apply biotechnology in improving seafood processing. Research should be conducted to develop and improve technologies for all these applications.

CONSERVING GENETIC RESOURCES

The preservation and enhancement of biodiversity in natural systems is an important Federal priority. Therefore, the Federal Government should encourage and support programs to maintain and enhance biodiversity in aquatic systems through cultivation and stocking of aquatic species. Biotechnology can be employed in two ways to conserve genetic resources of aquatic species.

First, the tools of biotechnology can be used to identify and characterize important aquatic germplasm, including endangered species. Genomes of aquatic species can be analysed and characterized, and quantitative trait loci identified. Second, biotechnology can be applied to improve understanding of the molecular basis of gene regulation and expression as well as sex determination and thereby improve methods for defining species, stocks, and populations. Approaches include developing marker-assisted selection technologies, improving precision and efficiency of transgenic techniques, and improving technologies for the cryopreservation of gametes and embryos. Ultimately, stocking certain areas with selected, cultivated species and strains could help maintain biodiversity in natural aquatic ecosystems.

ENHANCING BIOMEDICAL MODELS

Aquaculture has important purposes other than food production. Because they often adapt to extreme environments, marine organisms can provide unique models for research on biological and physiological processes.

Studies of the developmental, cellular, and molecular aspects of marine organisms as model systems will provide insights into the basis of disease mechanisms and pathogenesis in humans. By contrast, use of mammalian organisms as a basis for the development of some types of human disease models may be neither feasible nor cost effective.

Progress in this arena will require that sophisticated molecular biology technologies be adapted to marine organisms, in order to enhance understanding of their biological processes. For example, approaches for gene transfer into eggs have been developed for many terrestrial organisms, but not for most marine species. This technology is needed for analyses of gene regulatory systems and gene expression. In addition, methods need to be developed for culturing tissues from marine organisms. Cultured cell lines will provide opportunities for gene transfer and gene expression studies and enhance the usefulness of marine species as biomedical research models. This is an important research area deserving of Federal support.

HARMFUL EFFECTS OF GENETICALLY MODIFIED FOODS

Largely between 1997 and 1999, genetically modified (GM) food ingredients suddenly appeared in *2/3rds of all US processed foods*. This food alteration was fuelled by a single Supreme Court ruling. It allowed, for the first time, the patenting of life forms for commercialization. Since then thousands of applications for experimental genetically-modified (GM) organisms, including quite bizarre GMOs, have been filed with the US Patent Office alone, and many more abroad. Furthermore an economic war broke out to own equity in firms that legally claimed such patent rights or the means to control not only genetically modified organisms but vast reaches of human food supplies. This has been the behind-the-scenes and key factor for some of the largest and rapid agri-chemical firm mergers in history.

The merger of Pioneer Hi-Bed and Dupont (1997), Novartis AG and AstraZeneca PLC (2000), plus Dow's merger with Rohm and Haas (2001) are three prominent examples, Few consumers are aware this has been going on and is ever continuing. Yet if you recently ate soya sauce in a Chinese restaurant, munched popcorn in a movie theatre, or indulged in an occasional candy bar-you've undoubtedly ingested this new type of food. You may have, at the time, known exactly how much salt, fat and carbohydrates were in each of these foods because regulations mandate their labelling for dietary purposes. But you would not know if the bulk of these foods, and literally every cell had been genetically altered!

In just those three years, as much as 1/4 of all American agricultural lands or 70-80 million acres were quickly converted to raise genetically-modified (GM) food and crops. And in the race to increase GM crop production verses organics, the former is winning.

CORE PHILOSOPHICAL ISSUES

When Gandhi confronted British rule and Martin Luther King addressed those who disenfranchised Afro-Americans, each brought forth issues of morality and spirituality. They both challenged others to live up to the highest principles of humanity. With the issue of GM food technology, we should naturally do the same, and with great respect for both sides. It is not enough to list fifty or more harmful effects but we need to also address moral, spiritual and especially worldview issues. Here the stakes are incredibly huge. The philosophical issues involving GMOs, why this technology represents the impregnation of a mechanical worldview, a death-centred vision of nature that is greatlyt accelerating the death of species on earth.

FROM HYBRIDIZATION TO GMOS

Another challenging phenomenon to face in our modern world is that of hybridization. It seems to have worked so very successfully in some commercial realms, and as a major application of Gregor Mendel's revolutionary Gene Theory. Mendel offered a logical extension of the larger mechanical worldview. Just as we create factory assembly lines for manufacturing inanimate products, why can't we also manufacture living organisms, and using the same or similar principles? Why not take this assembly-line process to the next logical and progressive level?

What's wrong then with the "advance" of genetic engineering? No doubt, with hybridizations conscious life is manipulated. But living organisms continue to make some primary genetic decisions amid limited selections. We can understand this with an analogy. There is an immense difference between being a matchmaker and inviting two people to a dinner party, to meet and see if they are compatible. This differs essentially from forcing their meeting and union or a violent date rape. The former act may be divine, and the latter considered criminal. The implication is that biotechnology involves vital moral issues in regard to the whole of life in nature.

With biotechnology, roses are no longer crossed with just roses. They are mated with pigs, tomatoes with oak trees, fish with asses, butterflies with worms, orchids with snakes. The technology that makes this all possible is called biolistics-a gunshot-like violence that pierces the nuclear membrane of cells. This essentially violates not just the core chambers of life (physically crossing nuclear membranes) but the conscious-choice principle that is part of living nature's essence. Some also compare it to the violent crossing of territorial borders of countries, subduing inhabitants against their will.

What will happen if this technology is allowed to spread? Fifty years ago few predicted that chemical pollution would cause so much vast environmental harm. Now nearly 1/3rd of all species are threatened with extinction (and up to half of all plant species and half of all mammals). Few also knew that cancer rates would skyrocket during this same period. Nowadays approximately 41per cent on average of Americans can expect cancer in their lifetime.

ALARM SIGNALS

No one has a crystal ball to see future consequences of the overall GMO technology. Nevertheless, there are silent alarm signals like the early death of canaries in a mine shaft. There is, for example, growing evidence that the wholesale disappearance of bees relates directly to the appearance of ever more GM pollen. If we understand certain philosophical issues about the 17th century's worldview, the potential harm of GMOs actually can potentially far outweigh that of chemical pollution.

This is because chemistry deals mostly with things altered by fire (and then no longer alive, isolated in laboratories-and not infecting living terrains in self-reproducible ways). Thus a farmer may use a chemical for many decades, and then let the land lie fallow to convert it back to organic farming. This is because the chemicals tend to break down into natural substances over time, Genetic pollution, however, can alter the oil's life forever!

Farmers who view their land as their primary financial asset have reason to heed this warning. They need to be alarmed by evidence that genetically-modified soil bacteria contamination can arise. This is more than just possible, given the numerous (1600 or more) distinct microorganisms that can be found in a single teaspoon of soil. If that soil contamination remains permanently, the consequences can be catastrophic. Someday the public may blacklist precisely those farms that have once planted GM crops. No one has put up any warning signs on product packaging for farmers, including those who now own 1/4 of all agricultural tracks in the US. Furthermore, the spreading potential impact on all ecosystems is profound.

"Our way of life is likely to be more fundamentally transformed in the next several decades than in the previous one thousand years...Tens of thousands of novel transgenic bacteria, viruses, plants and animals could be released into the Earth's ecosystems...Some of those releases, however, could wreak havoc with the planet's biospheres."

In short these processes involve unparalleled risks. Voices from many sides echo this view. Contradicting safety claims, no major insurance company has been willing to limit risks, or insure bio-engineered agricultural products. The reason given is the high level of unpredictable consequences. Over eight hundred scientists from 84 countries have signed The World Scientist open letter to all governments calling for a ban on the patenting of life-forms and emphasizing the very grave hazards of GMOs, genetically-modified seeds and GM foods. This was submitted to the UN, World Trade Organization and US Congress. The Union of Concerned Scientists (a 1000 plus member organization with many Nobel Laureates) has similarly expressed its scientific reservations. The prestigious medical journal, Lancet, published an article on the research of Arpad Pusztai showing potentially significant harms, and to instill debate. Britain's Medical Association (the equivalent of the AMA and with over a 100,000 physicians) called for an outright banning of genetically-modified foods and labelling the same in countries where they still exist. In a gathering of political representatives from over 130 nations, drafting the Cartagena Protocol on Biosafety, approximately 95per cent insisted on new precautionary approaches. The National Academy of Science report on genetically-modified products urged greater scrutiny and assessments. Prominent FDA scientists have repeatedly expressed profound fears and reservations but their voices were muted not due to cogent scientific reasons but intense political pressure from the Bush administration in its efforts to buttress and promote the profit-potentials of a nascent biotech industry.

To counterbalance this, industry-employed scientists have signed a statement in favour of genetically-modified foods. But are any of these scientists impartial? Writes the New York Times (about a similar crisis involving genetic engineering and medical applications).

"Academic scientists who lack industry ties have become as rare as giant pandas in the wild...lawmakers, bioethics experts and federal regulators are troubled that so many researchers have a financial stake [via stock options or patent participation] ...The fear is that the lure of profit could colour scientific integrity, promoting researchers to withhold information about potentially dangerous side-effects."

Looked at from outside of commercial interests, perils of genetically modified foods and organisms are multi-dimensional. They include the creation of new "transgenic" life forms-organisms that cross unnatural gene lines (such as tomato seed genes crossed with fish genes)-and that have unpredictable behaviour or replicate themselves out of control in the wild. This can happen, without warning, *inside of our bodies* creating an unpredictable chain reaction. A four-year study at the University of Jena in Germany conducted by Hans-Hinrich Kaatz revealed that bees ingesting pollen from transgenic rapeseed had bacteria in their gut with modified genes. This is called a "horizontal gene transfer." Commonly found bacteria and microorganisms in the human gut help maintain a healthy intestinal flora. These, however, can be mutated.

Mutations may also be able to travel internally to other cells, tissue systems and organs throughout the human body. Not to be underestimated, the potential domino effect of internal and external genetic pollution can make the substance of science-fiction horror movies become terrible realities in the future. The same is true for the bacteria that maintain the health of our soil- and are vitally necessary for all forms of farming-in fact for human sustenance and survival.

Without factoring in biotechnology, milder forms of controlling nature have gravitated towards restrictive monocropping. In the past 50 years, this underlies the disappearance of approximately 95per cent of many native grains, beans, nuts, fruits, and vegetable varieties in the United States, India, and Argentina among other nations (and on average 75per cent worldwide). Genetically-modified monoculture, however, can lead to yet greater harm. Monsanto, for example, had set a goal of converting *100per cent* of all US soy crops to Roundup Ready strains by the year 2000. If this plan were effected, it would have threatened the biodiversity and resilience of all future soy farming practices. Monsanto laid out similar strategies for corn, cotton, wheat and rice. This represents a deepest misunderstanding of how seeds interact, adapt and change with the *living* world of nature.

One need only look at agricultural history-at the havoc created by the Irish potato blight, the Mediterranean fruit fly epidemic in California, the regional citrus canker attacks in the Southeast, and the 1970's US corn leaf blight. In the latter case, 15per cent of US corn production was quickly destroyed. Had weather changes not quickly ensued, most all crops would have been laid waste because a fungus attached their cytoplasm universally. The deeper reason this happened was that approximately 80per cent of US corn had been standardized (devitalized/mechanized) to help farmers crossbreed-and by a method akin to those used in current genetic engineering. The uniformity of plants then allowed a single fungus to spread, and within four months to destroy crops in 581 counties and 28 states in the US. According to J. Browning of Iowa State University: "Such an extensive, homogeneous acreage of plants... is like a tinder-dry prairie waiting for a spark to ignite it."

The homogeneity is unnatural, a byproduct again of deadening nature's creativity in the attempt to mechanize, to grasp absolute control, and of what ultimately yields not control but wholesale disaster. Europeans seem more sensitive than Americans to such approaches, given the analogous metaphor of German eugenics.

HISTORICAL SYNOPSIS

Overall the "biotech revolution" that is presently trying to overturn 12,000 years of traditional and sustainable agriculture was launched in the summer of 1980 in the US. This was the result of a little-known US Supreme Court decision *Diamond vs. Chakrabarty* where the highest court decided that biological life could be legally patentable.

Ananda Mohan Chakrabarty, a microbiologist and employee of General Electric (GE), developed at the time a type of bacteria that could ingest oil. GE rushed to apply for a patent in 1971. After several years of review, the US Patent and Trademark Office (PTO) turned down the request under the traditional doctrine that life forms are not patentable. Jeremy Rifkin's organization, the Peoples Business Commission, filed the only brief in support of the ruling. GE later sued and won an overturning of the PTO ruling. This gave the go ahead to further bacterial gmo research throughout the 1970's. Then in 1983 the first genetically-modified plant, an anti-biotic resistant tobacco was introduced. Field trials then began in 1985, and the EPA approved the very first release of a GMO crop in 1986. This was a herbicide-resistant tobacco. All of this went forward due to a regulatory green light as in 1985 the PTO also decided the Chakrabarty ruling could be further extended to all plants and seeds, or the entire plant kingdom. It then took another decade before the first genetically-altered crop was commercially introduced. This was the famous delayed-ripening "Flavr-savr" tomato approved by the FDA on May 18, 1994.

The tomato was fed in laboratory trials to mice who, normally relishing tomatoes, refused to eat these lab-creations and had to be force-fed by tubes. Several developed stomach lesions and seven of the forty mice died within two weeks. Without further safety testing the tomato was FDA approved for commercialization. Fortunately, it ended up as a production and commercial failure, and was ultimately abandoned in 1996. This was the same year Calgene, the producer, began to be bought out by Monsanto. During this period also, and scouring the world for valuable genetic materials, W.R. Grace applied for and was granted fifty US patents on the neem tree in India. It even patented the indigenous knowledge of how to medicinally use the tree (what has since been called biopiracy). Also by the close of the 20th century, about a dozen of the major US crops-including corn, soy, potato, beets, papaya, squash, tomato and cotton-were approved for genetic modification.

Going a step further, on April 12, 1988, PTO issued its first patent on animal life forms (known as oncomice) to Harvard Professor Philip Leder and Timothy A. Stewart. This involved the creation of a transgenic mouse containing chicken and human genes. Since 1991 the PTO has controversially granted other patent rights involving human stem cells, and later human genes. A United States company, Biocyte was awarded a European patent on all umbilical cord cells from fetuses and newborn babies. The patent extended exclusive rights to use the cells without the permission of the donors. Finally the European Patent Office (EPO) received applications from Baylor University for the patenting of women who had been genetically altered to produce proteins in their mammary glands. Baylor essentially sought monopoly rights over the use of human mammary glands to manufacture pharmaceuticals. Other attempts have been made to patent cells of indigenous peoples in Panama, the Solomon Islands, and Papua New Guinea, among others.

Thus the groundbreaking Chakrabarty ruling evolved, and within little more than two decades from the patenting of tiny, almost invisible microbes, to allow the genetic modification of virtually all terrains of life on Earth.

Certain biotech companies then quickly, again with lightening speed, moved to utilize such patenting for the control of first and primarily seed stock, including buying up small seed companies and destroying their non-patented seeds. In the past few years, this has led to a near monopoly control of certain genetically modified commodities, especially soy, corn, and cotton (the latter used in processed foods when making cottonseed oil). As a result, between 70-75per cent of processed grocery products, as estimated by the Grocery Manufacturers of America, soon showed genetically-modified ingredients. Yet again without labelling, few consumers in the US were aware that any of this was pervasively occurring. Industry marketers found out that the more the public knew, the less they wanted to purchase GM foods. Thus a concerted effort was organized to convince regulators (or bribe them with revolving-door employment arrangements) not to require such labelling.

ECOLOGICAL EFFECTS OF HRCS

GENE FLOW

Are we altering the genetic structure of living things in the name of utility and profit?

Just as it occurs between traditionally improved crops and wild relatives, pollen mediates gene flow between GMCs and wild relatives or conspecifics despite all possible efforts to reduce it. Little is known about the long-term persistence of crop genes in wild populations or about the impact of fitness-related crop genes on the population dynamics of weedy relatives. The main concern with transgenes that confer significant biological advantages is that they may transform wild/weed plants into new or worse weeds.

Hybridization of HRCs with populations of free living relatives will make these plants increasingly difficult to control, especially if they are already recognized as agricultural weeds and if they acquire resistance to widely used herbicides. For example:

- Transgenic resistance to glufosinate can be passed on from Brassica napus to populations of weedy Brassica napa, and persist under natural conditions.
- In Europe there is a major concern about the possibility of pollen transfer of herbicide tolerant genes from *Brassica* oilseeds to *Brassica nigra* and *Sinapis arvensis*.

ECONOMIC AND AGRONOMIC IMPLICATIONS

The majority of GMCs have been engineered to repel unwanted plants or weeds.

Worldwide in 2000, transgenic herbicide resistant crops were planted on 74per cent of the 44.2 million hectares devoted to transgenic crops.

In North America, transgenic glufosinate resistant cultivars of canola and corn, and transgenic glyphosate resistant cultivars of soybean, corn, cotton, and canola are now commercially available. Bromoxynil resistant transgenic cotton has also been developed. The so-called Round-up ready soybeans are the most prevalent GMCs.

Transgenic herbicide resistance in crop plants simplifies chemically based weed management because it typically involves compounds that are active on a very broad spectrum of weed species. Post-emergence application timing for these materials fits well with reduced or zero-tillage production methods, which can conserve soil and reduce fuel and tillage costs. However, HRCs also have significant problems.

Editor's note: Sales of biotechnology products are reaching $60 billion/ year.

- Reliance on HRCs perpetuates the weed resistance problems and species shifts that are common to conventional herbicide based approaches.
- Herbicide resistance becomes more of a problem as the number of herbicide modes of action to which weeds are exposed becomes fewer and fewer, a trend that HRCs may exacerbate due to market forces.
- Given industry pressures to increase herbicide sales, acreage treated with broad-spectrum herbicides will expand, exacerbating the resistance problem. For example, it has been projected that the acreage treated with glyphosate will increase to nearly 150 million acres. Although glyphosate is considered less prone to weed resistance, the increased use of the herbicide will result in weed resistance, even if more slowly, as it has been already documented with Australian populations of annual ryegrass, quackgrass, birdsfoot trefoil and *Cirsium arvense*.
- Perhaps the greatest problem of using HRCs to solve weed problems is that they steer efforts away from crop diversification and help to maintain cropping systems dominated by one or two annual species. Crop diversification can
  — reduce the need for herbicides
  — improve soil and water quality
  — minimize requirements for synthetic nitrogen fertilizer
  — regulate insect pest and pathogen populations
  — increase crop yields and reduce yield variance.

Thus, to the extent that transgenic HRCs inhibit the adoption of diversified cropping systems that include rotational crops, cover crops and green manure, they hinder the development of sustainable agriculture.

AGRICULTURAL BIOTECHNOLOGIES IN DEVELOPING COUNTRIES

Here we provide a brief overview of the main kinds of agricultural biotechnologies that have been used in developing countries over the past 20 years and that should be covered in the e-mail conference. They are described separately, although in practice more than one may be used together in certain situations. Note, new biotechnologies that are still at the research level, be it in the laboratory or at the field trial stage, but which have not yet been applied (*i.e.* used for commercial production by farmers) in developing countries are not included. A short description of the different biotechnologies is provided below, indicating also what they are used for, the food and agricultural sectors involved and giving some examples of their applications in specific developing countries. Regarding the examples, their inclusion in the document does not imply that these applications have been a partial or complete success (or, conversely, that they have been any kind of a failure). Indeed, these are the kind of issues to be addressed by participants during this e-mail conference. Also, it should be kept in mind that, although not the subject of this e-mail conference, the pathway from a research development in the laboratory to its eventual application in the field (*e.g.* farmers cultivating a new genetically improved plant variety or using a new vaccine against an animal disease) can be long, resource-demanding and unsuccessful, so many biotechnologies of seemingly high promise at the experimental stage have had limited applications in developing countries so far.

As many of the biotechnologies are related to molecular biology and genetic material, some basic terminology is introduced here. Living things are made up of cells that are programmed by genetic material called DNA. A DNA molecule is made up of a long chain of nitrogen-containing bases. Only a small fraction of this DNA sequence typically makes up genes *i.e.* that code for proteins, which are molecules essential for the functioning of living cells, made up of chains of amino acids. The remaining and major share of the DNA represents non-coding sequences whose role is not yet clearly understood.

The genetic material is organized into sets of chromosomes (*e.g.* 5 pairs in Arabidopsis thaliana-a model plant species; 30 pairs in cattle), and the entire set is called the genome. In a diploid individual (*i.e.* where chromosomes are organized in pairs), there are two alleles of every gene-one from each parent-transmitted by gametes (reproductive cells) that are normally haploid (having just one of each of the pairs of chromosomes). A typical genome contains several thousand genes *e.g.* about 30,000 genes in grasses like rice and sorghum.

MOLECULAR MARKERS

Molecular markers are identifiable DNA sequences, found at specific locations of the genome, transmitted by standard Mendelian laws of inheritance from one generation to the next. They rely on a DNA assay and a range of different kinds of molecular marker systems exist, such as restriction fragment length polymorphisms (RFLPs), random amplified polymorphic DNAs (RAPDs), amplified fragment length polymorphisms (AFLPs) and microsatellites. The technology has improved in the past decade and faster, cheaper systems like single nucleotide polymorphisms (SNPs) are increasingly being used. The different marker systems may vary in aspects such as their technical requirements, the amount of time, money and labour needed and the number of genetic markers that can be detected throughout the genome.

Molecular markers have been used in laboratories since the late 1970s and they are applied across all the food and agricultural sectors. They are very versatile and can be used for a variety of purposes.

Thus, they are used in genetic improvement, through so-called marker-assisted selection (MAS), where markers physically located beside (or, even, within) genes of interest (such as those affecting yield in maize) are used to select favourable variants of the genes. MAS is made possible by the development of molecular marker maps, where many markers of known location are interspersed at relatively short intervals throughout the genome, and the subsequent testing for statistical associations between marker variants and the traits of interest. Marker maps are now available for a wide range of economically important agricultural species. Progress in the field of genomics (the study of an organism's entire genome) has also provided much useful information for MAS, enabling in some cases markers to be used that are located within the genes of interest.

Molecular markers are also used to characterize and conserve genetic resources, where some of the approaches can be applied in each of the crop, forestry, livestock and fishery sectors (*e.g.* estimating the genetic relationships between populations within a species). Other uses again are more sector-specific, such as their utilization to identify duplicate accessions in crop genebanks; monitor effective population sizes (Ne) in capture fish populations or carry out biological studies (*e.g.* of mating systems, pollen movement and seed dispersal) in forest tree populations. They are also used in disease diagnosis, to characterize and detect pathogens in livestock, crops, forest trees, fish and food.

Molecular markers have been used in a number of developing countries. In livestock, for example, they have been used in four African countries for characterization of genetic resources and in eight Asian countries, where six used them for genetic distance studies and two for MAS. In Latin America and the Caribbean, most countries have used molecular techniques, primarily for characterization purposes, while their use has been limited in the Near and Middle East. In crops, several examples of new hybrids and varieties developed through MAS are available, and in progress, in different crops, such as pearl millet, rice and maize, and in several developing countries like Bangladesh, India and Thailand. Different centres of the Consultative Group on International Agricultural Research (CGIAR) have been working with partners in developing countries to accelerate plant breeding practices through MAS.

GENETIC MODIFICATION

A genetically modified organism (GMO) is an organism in which one or more genes (called transgenes) have been introduced into its genetic material from another organism. The genes may be from a different kingdom (*e.g.* a bacterial gene introduced into plant genetic material), a different species within the same kingdom or even from the same species. For example, so-called 'Bt crops' are crops containing genes derived from the soil bacterium Bacillus thuriengensis coding for proteins that are toxic to insect pests that feed on the crops. The issue of GMOs has been highly controversial over the past decade. Many countries have introduced specific frameworks to regulate their release and commercialization.

GM crops were first grown commercially in the mid 1990s. While the majority continues to be grown in developed countries, an increasing number of developing countries are reported to be cultivating them.

Recent estimates indicate that 10 developing countries planted over 50,000 hectares of GM crops in 2008 *i.e.* Argentina (21.0 million hectares), Brazil (15.8), India (7.6), China (3.8), Paraguay (2.7), South Africa (1.8), Uruguay (0.7), Bolivia (0.6), Philippines (0.4) and Mexico (0.1). For comparison, in 1997 the only developing countries reported were Argentina (1.4 million hectares), China (1.8) and Mexico (less than 0.1). Almost all GM crops grown commercially are genetically modified for one or both of two main traits: herbicide tolerance (63per cent of GM crops planted in 2008) or insect resistance (15per cent), *i.e.* Bt crops, while 22per cent have both traits. Commercial release of GM forest trees has been reported in one country, China. In 2002, approval was granted for the environmental release of two kinds of Bt trees, the European black poplar (Populus nigra) and the hybrid white poplar clone GM 741, together representing about 1.4 million plants on 300-500 hectares (FAO, 2004). Regarding GM livestock or fish, there has been no commercial release for food and agriculture purposes in any developing or developed country.

Although documentation is generally quite poor, use of genetically modified microorganisms (GMMs) in the agro-industry and animal feed sector is routine in developed countries and is also a reality in many developing countries. In the agro-industry sector, use of enzymes, *i.e.* proteins that catalyse specific chemical reactions, is important. Many of the enzymes used in the food industry are commonly produced using GMMs. For example, since the early 1990s, preparations containing chymosin (an enzyme used to curdle milk in the preliminary steps of cheese manufacture) derived from GM bacteria have been available commercially. Similarly, many colours, vitamins and essential amino acids used in the food industry are also from GMMs. In animal nutrition, feed additives such as amino acids and enzymes are widely used in developing countries. The greatest use is in pig and poultry production where, over the last decade, intensification has increased, further accelerating the demand for feed additives. For example, most grain-based livestock feeds are deficient in essential amino acids such as lysine, methionine and tryptophan and for high producing monogastric animals (pigs and poultry) these amino acids are added to diets to increase productivity. The amino acids in feed, L-lysine, DL-methionine, L-threonine and L-tryptophan, constitute over half of the total amino acid market. In India alone, the amino acid market amounts to about 5 million US dollars. The essential amino acids are produced in some cases by GM strains of Escherichia coli.

In the dairy industry, recombinant bovine somatotropin (rBST), a protein hormone from an Escherichia coli K-12 bacterium containing the cow somatotropin gene, has been used to increase milk production in a number of developing countries. Chauvet and Ochoa (1996) report that rBST was first used in Mexico in 1990 and has been sold in a number of other developing countries, including Brazil, Malaysia, South Africa and Zimbabwe.

CHROMOSOME NUMBER MANIPULATION

As mentioned earlier, genetic material is organized into sets of chromosomes and each plant and animal species has a characteristic number of chromosomes. Manipulation of whole sets of chromosomes is possible and is used for a range of different purposes in agriculture. For example, fish and shellfish have been extensively studied for manipulation of their chromosomes during early stages of development. Using relatively simple techniques such as cold and hot shocks it is possible to induce triploidy (*i.e.* with three sets of chromosomes), leading to the production of nearly completely sterile populations. Sterility may be desirable in conservation programs, where it can prevent introgression of escaped individuals from commercial stocks into natural populations. It may also be of interest in commercial fish operations, *e.g.* when developing hybrid stocks or to prevent the side effects of sexual maturation on carcass quality. As in fish, induction of sterility in crops may be desirable in certain breeding programmes, *e.g.* to produce seedless fruits, and one of the most rapid and cost-effective approaches is to create polyploids (*i.e.* with more than 2 complete sets of chromosomes), especially triploids. Triploid varieties have been produced in numerous fruit crops including most of the citrus fruits, acacias and the kiwifruit.

Another example of chromosomal number modification in fish is the production of haploid individuals after eggs are fertilized by sperm which do not contribute genetic material (a process called gynogenesis) or when normal sperm fertilize eggs whose DNA has been deactivated (a process called androgenesis). In both cases the haploid chromosomes can then be duplicated using shocks. The importance of gynogenesis/androgenesis is that it is possible to develop inbred individuals, which may be useful in fish breeding experiments aimed at producing clonal lines for detecting genomic regions affecting quantitative traits.

In crops, chromosome doubling is one of the most important technologies for the creation of fertile inter-specific hybrids (wide-crosses). Wide crossing involves hybridizing a crop variety with a distantly related plant from outside its normal sexually compatible gene pool. Its usual purpose is to obtain a plant that is virtually identical to the original crop, except for a few genes contributed by the distant relative. The technique has enabled breeders to access genetic variation beyond the normal reproductive barriers of their crops. For example, the New Rice for Africa (NERICA) hybrids are derived from crossing two species of cultivated rice, the African rice and the Asian rice, combining the high yields from the Asian rice with the ability of the African rice to thrive in harsh environments.

Wide-hybrid plants are often sterile so their seed cannot be propagated, due to differences between the sets of chromosomes inherited from genetically divergent parental species that prevent stable chromosome pairing during meiosis. However, if chromosome number is artificially doubled, the hybrid may be able to produce functional pollen and eggs and be fertile. Colchicine has been used for chromosome doubling in plants since the 1940s and has been applied to more than 50 plant species, including most important annual crops. More recently, several additional chromosome doubling agents, all of which act as inhibitors of mitotic cell division, have been used in plant breeding programmes. To date, with the help of chromosome doubling technology, hundreds of new varieties have been produced worldwide.

In crops and forest trees, chromosome doubling has also been used, as for fish, to generate 'doubled haploids'. The haploid plants can be produced using anther culture which involves the in vitro culture of immature anthers. As the pollen grains are haploid, the resulting pollen-derived plants are also haploid. Doubled haploid plants were first produced in the 1960s using colchicine and today, thermal shock or mannitol incubation can be used. They may also be produced from ovule culture. Breeders value doubled haploid plants because they are 100per cent homozygous so any recessive genes are readily apparent. The time required after a conventional hybridization to select pure lines carrying the required recombination of characters is thus drastically reduced. Since the 1970s, doubled haploid methods have been used to create new varieties of barley, wheat, rice, melon, pepper, tobacco, and several Brassicas. In the developing world, a major centre of such breeding work is China, where numerous haploid crops have been released and many more are being developed. By 2003, China was cultivating over 2 million hectares of doubled haploid varieties, the most important of which are rice, wheat, tobacco and peppers.

THE PROMISE OF AGRICULTURAL BIOTECHNOLOGY

Due in large part to scientific advances in crop breeding and farming techniques, world food production has doubled since 1960, and productivity from agricultural land and water usage has tripled. Today, the peoples of the world farm an area about the size of South America; without the scientific advances of the past 30 years, farmland equivalent to the entire Western hemisphere would be required.

But a dilemma lies ahead. The world's population is expected to double by the year 2030 to 12 billion, and it is not clear whether current food production is keeping pace with population growth. The question is how best to feed billions of additional people without destroying much of the planet in the process. The disappearance of tropical rain forests, wetlands, and other vital habitats will accelerate unless agriculture somehow becomes more productive and less taxing to the environment.

It seems certain that agricultural biotechnology will play a major role in resolving this dilemma. Biotechnology can be employed to improve the quality of seed grains; increase protein levels in forage crops; and instill in crops resistance to disease, insects, and viruses, as well as tolerance for droughts, floods, and extreme temperatures. In addition, biotechnology can make foods healthier and more nutritious. For example, tomatoes and other fruits and vegetables containing increased levels of certain nutrients, such as vitamins C and E and beta carotene, may help protect against risk of chronic diseases, such as some cancers and heart disease.

Although major genetic improvements have been made in crops, progress in conventional breeding programs has been slow. Moreover, most crops grown in the United States produce less than 50 percent of their genetic potential. These shortfalls in yield are due in large part to the inability of crops to tolerate or adapt to environmental stresses, pests, and disease. For example, some of the world's highest yields of potatoes are in Idaho under irrigation, but in 1993 both quality and yield were reduced severely because of cold, wet weather and widespread frost damage during June. And some of the world's best bread wheats and malting barleys are produced in the north-central states, but in 1993 the disease Fusarium (headblight of wheat and barley) caused an estimated $1 billion in damage. Major advances also have been made through conventional breeding and selection of livestock. Feed efficiency (1) for poultry and swine has been increased by nearly 50 percent. Milk production per cow has more than doubled since 1955. Diseases such as hog cholera and pests such as screwworm have been eradicated. Yet major diseases of livestock still go uncontrolled and some still go undiagnosed or even unrecognized. These and many other agricultural production hazards have defied traditional solutions such as classical breeding approaches. New research is needed to address these problems, in order to minimize risks; improve the financial stability of farms, rural communities, and agribusinesses; and assure the long-term competitiveness of U.S. agriculture in the world market. These advances must be accompanied by reduced dependency on pesticides and improved environmental sustainability.

Biotechnology can compress the time frame required to translate fundamental discoveries into applications. With improved technology and knowledge about agricultural organisms, processes, and ecosystems, opportunities will emerge to produce new and improved agricultural products in an environmentally sound manner. This chapter highlights five broad priorities in agricultural biotechnology research that merit attention by Federal agencies:

- Continue mapping and sequencing of animal/plant/microbial genomes to elucidate gene function and regulation and to facilitate the discovery of new genes as a prelude to gene modification.
- Determine biochemical and genetic control mechanisms of metabolic pathways in animals, plants, and microbes that may lead to products with novel food, pharmaceutical, and industrial uses.
- Extend understanding of the biochemical and molecular basis of growth and development including structural biology of plants and animals.
- Elucidate the molecular basis of interactions of plants and animals with their physical and biological environments, as a basis for improving the organisms' health and wellbeing.
- Enhance food safety assurance methodologies, such as rapid tests for identifying chemical and biological contaminants in food and water.

This chapter identifies, within each priority area, selected areas where additional research is needed to assure a steady supply of high-quality agricultural products at reasonable prices for decades to come. The report also highlights recent scientific advances involving Federally supported research.

CONTROL OF INSECT PESTS PREYING ON PLANTS

Insect pests cause major crop damage. Examples include *Phylloxera*, which has caused devastation in European and Californian vineyards, and *Psylla*, which has ravaged the apple and pear industry of the eastern United States. In this century, the United States has imported and released approximately 800 natural enemies of insect pests; about 40 percent are providing some level of biological control. However, this approach to insect pest management has its limits. Further progress will require increased knowledge of interactions between pests and their natural enemies, and, if necessary, appropriate genetic modifications.

The chemicals emitted by plants in response to feeding by insect pests serve as an attractant to the natural enemies of some insect pests. Corn leaves attacked by beet armyworm caterpillars release volatile compounds that attract wasps that are parasites of these armyworms. Moreover, these parasitic wasps can recognize and fly towards the source of volatile compounds produced in response to feeding by their caterpillar hosts; the wasps ignore the odours released from leaves damaged mechanically, such as by mowing. Expanded knowledge of the "information molecules" used by beneficial insects to find their prey on crops could lead to development of crops capable of producing stronger signals, to attract higher populations of beneficial insects. To exploit fully this approach to biological control, more research is needed on the biochemistry and molecular biology of insect pests and their natural enemies.

Due to insects' evolving resistance to pesticides, and the removal of many of these chemicals from the market, there are few effective means of controlling sucking insects, such as whiteflies, aphids, and leafhoppers. These same insects are among the most important vectors (carriers) of viruses, rickettsia-like bacteria, and mycoplasma-like organisms responsible for major plant diseases. New synthetic chemical compounds to be on the market in the next few years primarily target chewing insects. Biological methods to control sucking insects also are advancing at the molecular level. An example is the enzymatic preparation of synthetic pyrethroids.

The salivary glands of aphids, whiteflies, and leafhoppers contain bacterial symbionts that provide essential amino acids to their insect hosts. Only a few of these symbionts have been characterized. Research is needed on the basic biology, molecular biology, and production of these bacterial symbionts. If the bacterial symbionts could be disrupted in some way, either by manipulating the genes of the symbionts or by engineering an antimicrobial agent into a plant, then an innovative control mechanism could be developed.

BENEFICIAL MICROORGANISMS IN PLANTS

Plants selectively aid the growth of specific types of beneficial microorganisms. Some microorganisms, for instance, have been shown to provide growth factors for plants and protect plants against insect attack and infection. Information about the molecular signals and genetic controls of these associations can lead to the use of these microorganisms to improve the performance — and especially the consistency of performance— of important food and fibre crops, and even to enhance the effectiveness of the microbes themselves.

ANIMAL PESTS AND DISEASES

Work is underway to introduce new genes into animals in order to impart resistance to disease and to identify host cell traits that can be used to enhance resistance to disease agents. For instance, a specific receptor on pig intestinal cells is required for *E. coli* to bind to the cell. Strains that do not carry the receptor— or have the receptor deleted through the tools of biotechnology are less susceptible to the diarrheal disease caused by the bacterium. Other research on enteric rotavirus infections in pigs has identified specific cell membrane components required for attachment of the rotavirus to an intestinal cell. A complementary molecule can be constructed using biotechnological methods to saturate and block this receptor site. Porcine stress syndrome is an inherited condition in swine that can lead to meat that is of poor quality or unusable. This condition causes significant losses in the swine industry. At the same time, animals susceptible to this genetic disease have highly favourable muscling with low fat content. A candidate gene for this disease has been identified and research now is needed to develop a DNA probe for the trait, as a means to identify susceptible animals and eliminate them from breeding programs. This research will allow the development of swine with high-quality, low-fat meat but without the trait associated with the disease.

FOOD SAFETY

Enhance food safety assurance methodologies, such as rapid tests for identifying chemical and biological contaminants in food and water.

Food-and waterborne illnesses caused by microorganisms pathogenic to humans have been estimated to affect more than 80 million Americans and cost the U.S. economy over $40 billion annually. For example, some 9,000 Americans die each year from food-borne illnesses caused by microorganisms such as *E. coli* in meat and *Salmonella* in poultry. At present, virtually all food inspection is visual. Rapid, accurate, non-invasive tools of biotechnology can help improve the detection and control of food-borne human pathogens as well as chemical contaminants. Federal support is needed in this area to assure that these tools are developed in a timely manner.

MARINE BIOCHEMICAL PROCESSES AND BIOMOLECULAR MATERIALS

The Marine Processes objective involves the description of the dynamics of coastal waters off Southern Iberia, combined with chemical characterization, to accurately assess the primary productivity in shelf waters. CIMA's approach to these phenomena is based on modelling, analyses and observations both in situ and remote. The major topics covered are:

- Models and measurements of transfer phenomena in the ocean,
- Chemical dynamics in the ocean,
- Marine microbial dynamics and foodwebs.

Recent research has demonstrated that marine biochemical processes can be exploited to produce new biomaterials. For example, a corporation in Chicago is commercializing a new class of biodegradable polymers modelled on natural substances that form the organic matrices of mollusk shells. Equally exciting are the mechanisms used by marine diatoms, coccolithophorids, mollusks, and other marine invertebrates to generate elaborate mineralized structures on a nanometer scale (less than a billionth of a meter in size). Nanometer-scale structures can have unusual and useful properties.

Research that will enhance understanding and allow engineering of the processes for creating these bioceramics promises to revolutionize the manufacturing of medical implants, automotive parts, electronic devices, protective coatings, and other novel products. Biomaterials also hold promise for counteracting biofouling, which long has been recognized as an extensive and costly problem. Bacterial biofilms form slime layers that increase drag on moving ships, interfere with transfer on heat exchangers, block pipelines, and contribute to corrosion on metal surfaces. Bacterial and microalgal colonization of surfaces is accompanied by settlement of invertebrate larvae and algal spores, eventually leading to "hard fouling" and the need for costly cleaning.

The most effective anti-fouling coatings have utilized toxic chemicals, such as copper and organotins. There is an urgent need for Non-toxic biofouling control strategies, due to heightened recognition of the impact that toxic coatings can have on the environment. Research is needed on the attachment mechanisms of marine organisms and the natural products they employ to prevent fouling of their own surfaces. Molecular approaches to characterizing biofilm structure and development offer considerable potential for finding novel biofouling prevention strategies. It is now possible to determine the genes and pathways involved in regulation and synthesis of bacterial adhesive polymers.

Considerable progress has been made in understanding the nature and expression of surface polymers produced by microorganisms such as the nitrogen-fixing *Rhizobium* species and the opportunistic pathogen *Pseudomonas aeruginosa*. Similar approaches can been applied to marine biofilm bacteria, to find the genetic determinants of adhesive production and the environmental factors that regulate synthesis. Molecular biology techniques also can be applied to determine the basis of natural antifouling mechanisms. Many marine plants and animals remain free of attached bacteria, either because they produce repelling compounds or because their surface structure neutralizes bacterial adhesives. A product generated by the seagrass *Zostra marina* (eelgrass), for example, is an effective agent for preventing fouling by bacteria, algal spores, and a variety of hard-fouling barnacles and tube worms. Molecular characterization of natural fouling resistance could provide new strategies for fouling control. Potential applications include prevention of fouling in industrial pipelines or heat exchangers, improved design of trickling filters or aquaculture circulation systems, and control of biofilm infections of medical implants and prosthetic devices.

BIOMONITORS

Marine organisms can provide the basis for development of biosensors, bio-indicators, and diagnostic devices for medicine, aquaculture, and environmental monitoring. One type of biosensor employs the enzymes responsible for bioluminescence. The *lux* genes, which encode these enzymes, have been cloned from marine bacteria such as *Vibrio fischeri* and transferred successfully to a variety of plants and other bacteria. The *lux* genes typically are inserted into a gene sequence, or operon, that is functional only when stimulated by a defined environmental feature. The enzymes responsible for toluene degradation, for example, are synthesized only in the presence of toluene. When *lux* genes are inserted into a toluene operon, the engineered bacterium glows yellow-green in the presence of toluene. This genetically engineered system "reports" that biodegradation of a specific chemical, in this case toluene, is proceeding.

Another type of biomonitor that holds great promise is the gene probe, which can be used to identify organisms that pose health hazards or may be useful in research. Specific gene probes can be employed, for example, to detect human pathogens in seafood and recreational waters; fish pathogens in aquaculture systems; microorganisms capable of mediating desired chemical transformations (*e.g.*, toxic chemical degradation, CO_2 assimilation, metal reduction); and specific fish stocks in fish migration and recruitment studies.

BIOPESTICIDES

Natural marine products have the potential to replace chemical pesticides and other agents used to maximize crop yields and growth.

Continued Federal support for RandD in this area is likely to result in useful natural pesticides that would provide greater specificity and fewer harmful side effects than do conventional synthetic agents. Current U.S. expenditures for all pesticides amount to $47 billion annually; by the year 2000, biopesticides from marine and other sources are expected to capture an estimated 10 percent of this market.An example of a marine biopesticide in use today is Padan*TM*, which was developed from a bait worm's toxin known to ancient Japanese fishermen. This natural pesticide has demonstrated activity against larvae of the rice stem borer, the rice plant skipper, and the citrus leaf miner, among other pests. More recently, scientists in Montana discovered novel compounds in marine algae and marine sponges containing symbiotic microorganisms. These compounds promoted growth and stimulated germination and increased root and coleoptile lengths in test plants.

Several sponge and nudibranch species produce terpenes, a broad class of aromatic compounds used in solvents and perfumes and known to deter feeding by fish. Extracts derived from these same sponge and nudibranch species also demonstrated powerful insecticidal activity against two species, grasshoppers and the tobacco hornworm.

NEW AND IMPROVED PROCESSES FROM THE SEAS

Priority: Develop bioremediation strategies for application in the world's coastal oceans, where multiple uses — including wastewater disposal, recreation, fishing, and aquaculture — demand prevention and remediation of pollution; and develop bioprocessing strategies for improving sustainable industrial processes.

BIOREMEDIATION

Bioremediation shows great promise for addressing problems in marine environments and in aquaculture. These problems include catastrophic spills of oil in harbours and shipping lanes and around oil platforms; movement of toxic chemicals from land, through estuaries, into the coastal oceans; disposal of sewage sludge, bilge waste, and chemical process wastes; reclamation of minerals, such as manganese; and management of aquaculture and seafood processing waste.

The full potential for marine organisms and processes to contribute new waste treatment and site remediation technologies cannot be realized without enhanced understanding of the unique conditions in marine environments. For example, oxidation-reduction (redox) states can fluctuate in coastal and estuarine sediments. The impact of changing redox conditions on biodegradation of environmental contaminants must be understood before waste management and remediation strategies and predictive models can be developed for contaminated sediments.

BIOPROCESSING

The emerging discipline of bioprocess engineering involves the application of biological science in manufacturing, to produce products such as biopharmaceuticals and natural bioactive agents. Bioprocess engineering requires an understanding of the biological system employed (such as a marine organism), isolation and purification of a product, and translation of the product into a stable, efficacious, and convenient form.

An emerging area of interest is the potential of marine bacteria and fungi to produce unusual chemical structures with no parallels in terrestrial organisms. Small-scale studies have begun to indicate the richness of marine microorganisms as sources for novel lead structures.

UNDERSTANDING AND CONSERVING THE SEAS

Priority: Improve understanding of microbial physiology, genetics, biochemistry, and ecology in order to provide model systems for research and production systems for commerce, and to contribute to understanding and conservation of the seas.

Scientists have a powerful new array of sampling devices and measuring instruments that will accelerate greatly the acquisition of knowledge about ocean resources and foster their wise use.

These technologies include manned deep-sea submersibles, remotely operated vehicles, geosynchronous satellites, sophisticated acoustic measuring devices, pressure-retaining deep-sea samplers, geographic information systems, real-time flow cytometry, PCR and biomonitoring techniques, computerized databases, and other forms of information exchange and analysis.

These tools should be exploited to accelerate the discovery of unknown marine microorganisms and to expand understanding of known varieties.

Federal support for this research is essential, because only then will sufficient information be acquired to assure that practical applications will result. As new life forms and processes become known, and as understanding of them grows, marine biotechnology will make significant contributions to the nation's social and economic wellbeing.

THE ECOLOGICAL IMPACTS OF AGRICULTURAL BIOTECHNOLOGY

Transnational corporations (TNCs) such as Monsanto, DuPont, and Novartis, the main proponents of biotechnology, argue that carefully planned introduction of these crops should reduce or even eliminate the enormous crop losses due to weeds, insect pests, and pathogens. In fact, they argue that the use of such crops will have added beneficial effects on the environment by significantly reducing the use of agrochemicals.13 However, ecological theory predicts that as long as transgenic crops follow closely the pesticide paradigm prevalent in modern agriculture, such biotechnological products will do nothing but reinforce the pesticide treadmill in agroecosystems, thus legitimizing the concerns that many environmentalists and some scientists have expressed regarding the possible environmental risks of genetically engineered organisms. In fact, there are several widely accepted environmental drawbacks associated with the rapid deployment and widespread commercialization of such crops in large monocultures.

Biotechnology may someday be considered a safe agricultural tool but studies suggest it may have harmful ecological consequences, such as:

- spreading genetically-engineered genes to indigenous plants
- increasing toxicity, which may move through the food chain
- disrupting nature's system of pest control
- creating new weeds or virus strains.

The cassava makes up part of the diet of nearly 600 million people worldwide. By inserting a bacterial version of the gene for starch production, scientists have come up with a super-sized cassava. Photo:

David Monniaux.Transgenic crops (GMCs: genetically modified crops), main products of agricultural biotechnology, are increasingly becoming a dominant feature of the agricultural landscapes of the USA and other countries such as China, Argentina, Mexico and Canada.

Nearly half of American farms grow GMCs (genetically modified crops).

- Worldwide, the areas planted to transgenic crops jumped more than twenty-fold in the past six years, from 3 million hectares in 1996 to nearly 44.2 million hectares in 2000.
- In the USA, Argentina and Canada, over half of the average for major crops such as soybean, corn and canola are planted in transgenic varieties.

- Herbicide resistant crops (HRCs) and insect resistant crops (Bt crops) accounted respectively for 59 and 15 percent of the total global area of all transgenic crops in 2000.

Big business claims GMCs will reduce the use of chemical pesticides.

Transnational corporations (TNCs) such as Monsanto, DuPont, and Novartis, the main proponents of biotechnology, argue that carefully planned introduction of these crops should reduce or even eliminate the enormous crop losses due to weeds, insect pests, and pathogens. In fact, they argue that the use of such crops will have added beneficial effects on the environment by significantly reducing the use of agro chemicals. However, ecological theory predicts that as long as transgenic crops follow closely the pesticide paradigm prevalent in modern agriculture, such biotechnological products will do nothing but reinforce the pesticide treadmill in agro ecosystems, thus legitimizing the concerns that many environmentalists and some scientists have expressed regarding the possible environmental risks of genetically engineered organisms. In fact, there are several widely accepted environmental drawbacks associated with the rapid deployment and widespread commercialization of such crops in large monocultures, including:

Pests show rapid evolution in resisting the pesticide properties of GMCs.

Toxic buildup in GMCs harms useful insects.

- The spread of transgenes to related weeds or conspecifics via crop-weed hybridization
- Reduction of the fitness of non-target organisms through the acquisition of transgenic traits via hybridization
- The rapid evolution of resistance of insect pests such as Lepidoptera to Bt
- Accumulation of the insecticidal Bt toxin, which remains active in the soil after the crop is ploughed under and binds tightly to clays and humic acids;
- Disruption of natural control of insect pests through intertrophic-level effects of the Bt toxin on predators
- Unanticipated effects on non-target herbivorous insects (*i.e.*, monarch butterflies) through deposition of transgenic pollen on foliage of surrounding wild vegetation
- Vector-mediated horizontal gene transfer and recombination to create new pathogenic organisms.

This paper will focus on the known effects of the two dominant types of GMCs: herbicide resistant crops (HRCs) and insect resistant crops (Bt).

BIOTECHNOLOGY, AGRODIVERSITY AND FARMERS' OPTIONS

Monoculture, or farming only one crop, can lead to economic disaster and hunger. The spread of transgenic crops threatens crop diversity by promoting monocultures which leads to environmental simplification and genetic erosion. History has repeatedly shown that uniformity characterizing agricultural areas sown to a smaller number of varieties is a source of increased risk for farmers, as the genetically homogeneous fields may be more vulnerable to disease and pest attack. Farmers have many choices, other than biotechnology, that work.

Several people think that HRCs and Bt crops have been a poor choice of traits to feature the technology, given predicted environmental problems and the issue of resistance evolution. In fact, there is enough evidence to suggest that both these types of crops are not really needed to address the problems they were designed to solve. On the contrary, they tend to reduce the pest management options available to farmers. There are many alternative approaches, (*e.g.*, rotations, polycultures, cover crops, biological control, etc.) that farmers can use to effectively regulate the insect and weed populations that are being targeted by the biotechnology industry. To the extent that transgenic crops further entrench the current monocultural system, they impede farmers from using a plethora of alternative methods.

ECOLOGICAL RISKS OF BT CROPS

The number of crops engineered for insect resistance is on the rise.

Based on the fact that more than 500 species of pests have already evolved resistance to conventional insecticides, pests can also evolve resistance to Bt toxins present in transgenic crops. No one questions if Bt resistance will develop, the question is now how fast it will develop.

Susceptibility to Bt toxins can therefore be viewed as a natural resource that could be quickly depleted by inappropriate use of Bt crops. However, cautiously restricted use of these crops should substantially delay the evolution of resistance. But is cautious use of Bt crops possible given commercial pressures that have resulted in a rapid rollout of Bt crops reaching 8.2 million hectares worldwide in 2000?

Organic farmers can produce a significant yield without insecticides.

The refuge strategy of setting aside 20-30per cent of land to non-Bt crops to delay resistance is very difficult to implement regionally. Data from the Midwest shows that Bt corn saves on some insecticide use and yields are 2.4 Bu/acre higher than conventional corn but only under high European corn borer infestations (USDA 1999). On the other hand organic corn growers use no insecticides and obtain yields (4.8-9 t/ha) similar or slightly higher than conventional farmers (5.0-7.1 t/ha). GMCs may have unintended victims, such as the monarch butterfly or lacewing.

EFFECTS ON THE SOIL ECOSYSTEM

The possibilities for soil biota to be exposed to transgenic products are very high. The little research conducted in this area has already demonstrated:

Toxins from GMCs remain active in the soil, decreasing soil fertility.

- There is long term persistence of insecticidal products (Bt and proteinase inhibitors) in soil.
- The insecticidal toxin produced by Bacillus thuringiensis subsp. kurskatki remain active in the soil, where it binds rapidly and tightly to clays and humic acids.
- The bound toxin retains its insecticidal properties and is protected against microbial degradation by being bound to soil particles, persisting in various soils for at least 234 days.
- The presence of the toxin in exudates from Bt corn and verified that it was active in an insecticidal bioassay using larvae of the tobacco hornworm.

Given the persistence and the possible presence of exudates, there is potential for prolonged exposure of the microbial and invertebrate community to such toxins, and therefore studies should evaluate the effects of transgenic plants on both microbial and invertebrate communities and the ecological processes they mediate. If transgenic crops substantially alter soil biota and affect processes such as soil organic matter decomposition and mineralization, this would be of serious concern to organic farmers and most poor farmers in the developing world. These farmers cannot purchase or don't want to use expensive chemical fertilizers. They rely instead on local residues, organic matter and especially soil organisms for soil fertility (*e.g.*, key invertebrate, fungal or bacterial species) which can be affected by the soil bound toxin. Soil fertility could be dramatically reduced if crop leachates inhibit the activity of the soil biota and slow down natural rates of decomposition and nutrient release.

GENERAL CONCLUSIONS AND RECOMMENDATIONS

The available independently generated scientific information suggests that The long-term impacts of GMCs are not yet known.

- The massive use of transgenic crops poses substantial potential risks from an ecological point of view *the ecological effects are not limited to pest resistance and creation of new weeds or virus strains
- Transgenic crops can produce environmental toxins that move through the food chain and also may end up in the soil and water affecting invertebrates and probably ecological processes such as nutrient cycling
- No one can really predict the long-term impacts that will result from such massive deployment of such crops.

Not enough research has been done to evaluate the environmental and health risks of transgenic crops, an unfortunate trend. Most scientists feel that such knowledge is crucial to have before biotechnological innovations are implemented. There is a clear need to further assess the severity, magnitude and scope of risks associated with the massive field deployment of transgenic crops. Much of the evaluation of risks must move beyond comparing GMC fields and conventionally managed systems to include alternative cropping systems featuring crop diversity and low-external input approaches. This will allow real risk/benefit analysis of transgenic crops in relation to known and effective alternatives.

The loss of agricultural diversity may lead to disaster in developing countries. Moreover, the large-scale landscape homogenization with transgenic crops will exacerbate the ecological problems already associated with monoculture agriculture. Unquestioned expansion of this technology into developing countries may not be wise or desirable. There is strength in the agricultural diversity of many of these countries, and it should not be inhibited or reduced by extensive monoculture, especially when consequences of doing so results in serious social and environmental problems. The repeated use of transgenic crops in an area may result in cumulative effects such as those resulting from the buildup of toxins in soils. For this reason, risk assessment studies not only have to be of an ecological nature in order to capture effects on ecosystem processes, but also of sufficient duration so that probable accumulative effects can be detected. The application of multiple diagnostic methods will provide the most sensitive and comprehensive assessment of the potential ecological impact of transgenic crops.

BIOTECHNOLOGY-BASED DIAGNOSTICS

Applications of biotechnology for diagnostic purposes are important in crops, forest trees, livestock and fish as well as for food safety purposes. Two main kinds of methods are used, those based on the enzyme-linked immunosorbent assay (ELISA) and those based on the polymerase chain reaction (PCR).

Elisa systems are antibody-based techniques for the determination of the presence and quantity of specific molecules in a mixed sample. They are used in a range of formats, both for the detection of pathogens and for detection of antibodies produced by the host as a response to the pathogens, and a range of commercial kits are available, *e.g.* to detect fish and shrimp pathogens. Some of the ELISA-based methods use monoclonal antibodies, produced by a cell line that is both immortal and able to produce highly specific antibodies, or polyclonal antibodies, produced by many cell lines. In livestock, ELISAs form the large majority of prescribed tests for the OIE notifiable animal diseases, and many diagnostic kits are available in developing countries.

The PCR-based methods rely on the fact that each species of pathogen carries a unique DNA or RNA sequence that can be used to identify it. PCR allows production of a large quantity of a desired DNA from a complex mixture of heterogeneous sequences. It can amplify a selected region of 50 to several thousand DNA base pairs into billions of copies. After amplification, the target DNA can be identified using techniques such as gel electrophoresis or hybridization with a labelled nucleic acid (a probe). Real time PCR (or quantitative PCR) enables quantification of DNA or RNA present in a sample. The genomes of some viruses, such as the influenza A virus, are made of RNA instead of DNA, and to identify RNA from these viruses a complementary DNA (cDNA) copy of the RNA is first synthesized using an enzyme called reverse transcriptase. The cDNA then acts as the template to be amplified by PCR. This method is called reverse transcriptase PCR (RT-PCR).

PCR-based techniques offer high sensitivity and specificity and diagnostic kits allow the rapid screening of the virus or bacteria and have a direct use in situations where individuals show no antibody response after infection. For example, molluscs do not produce antibodies, and therefore antibody-based diagnostic tests are limited in their application to pathogen detection in these species. In fisheries, PCR-related tools are increasingly being used in developing countries, although they require detailed knowledge of the genomics of the pathogen itself and extensive validation in practice.

In livestock, public sector production of diagnostic kits for animal diseases in Asia and Latin America can be found in Brazil, Chile, China, India, Mexico and Thailand. Research capabilities for development, standardization and validation of diagnostic methods are also well advanced in these countries. PCR-based diagnostics are increasingly being employed in developing countries to back up findings from serological analyses. However, their use is largely restricted to laboratories of research institutions and universities and to the central and regional diagnostic laboratories run by governments. In aquaculture, there are some highly integrated companies operating in developing countries (*e.g.* in shrimp production) and these companies commonly use PCR-based diagnostic systems, where the analyses are either carried out by laboratories of the companies themselves or are outsourced to specialized private laboratories.

Biotechnology-based diagnostics are also important in food analysis. Many of the classical food microbiological methods used in the past were culture-based, with microorganisms grown on agar plates and detected through biochemical identification. These methods are often tedious, labour-intensive and slow. Genetic based diagnostic and identification systems can greatly enhance the specificity, sensitivity and speed of microbial testing. Molecular typing methodologies, commonly involving PCR, ribotyping (a method to determine homologies and differences between bacteria at the species or subspecies/strain level, using RFLP analysis of ribosomal RNA genes) and pulsed-field gel electrophoresis (a method of separating large DNA molecules on agarose gels), can be used to characterize and monitor the presence of spoilage flora (microbes causing food to become unfit for eating), normal flora and microflora in foods. RAPD or AFLP molecular marker systems can also be used for the comparison of genetic differences among species, subspecies and strains, depending on the reaction conditions used. The use of combinations of these technologies and other genetic tests allows the characterization and identification of organisms at the genus, species, subspecies and even strain levels, thereby making it possible to pinpoint sources of food contamination, to trace microorganisms throughout the food chain or to identify the causal agents of food-borne illnesses.

DEVELOPMENT OF VACCINES USING BIOTECHNOLOGIES

Immunization can be one of the most effective means of preventing and hence managing animal diseases. In general, vaccines offer considerable benefits for comparative low cost, a primary consideration for developing countries. In addition, development of good vaccines for important infectious diseases can lead to reduced use of antibiotics, which is an important issue in developing countries.As described by Kurath (2008), biotechnology has been used extensively in the development of vaccines for aquaculture, and is applied at each of the three main stages of vaccine development *i.e.* a) identification of potential antigen candidates that might be effective in vaccines (where an antigen is a molecule, usually a protein foreign to the fish, which elicits an immune response on first exposure to the immune system by stimulating the production of antibodies specific to its various antigenic determinants. During subsequent exposures, the antigen is bound and inactivated by these antibodies) b) construction of a new candidate vaccine (where biotechnology tools can be used to produce different kinds of vaccines such as DNA vaccines, recombinant vaccines or modified live recombinant viruses. For example, a DNA vaccine is a circular DNA plasmid containing a gene for a protective antigenic protein from a pathogen of interest and c assessment of candidate vaccine efficacy, its mode of action and the host response (where *e.g.* quantitative RT-PCR can be used to examine the expression of fish genes related to immune responses).

Of the countries that responded to a recent OIE survey, 4 out of 23 and 7 out of 14 African and Asian countries respectively indicated that they produce or use animal vaccines derived from biotechnology, including experimental use as well as commercial release.

REPRODUCTIVE BIOTECHNOLOGIES

A number of reproductive biotechnologies have been applied in developing countries to influence the number (and sex) of offspring from given individuals in fish and livestock populations.

ARTIFICIAL INSEMINATION

In artificial insemination (AI), semen is collected from donor male animals, diluted in suitable diluents and manually inseminated into the female reproductive tract during oestrus (heat), to achieve pregnancy. The semen can be fresh or preserved in liquid nitrogen and then thawed. Efficiency of AI can be increased by monitoring progesterone levels, *e.g.* using ELISA, to identify non-pregnant females, and/or by oestrus synchronization, where females are treated with hormones to being them into oestrus at a desired time.

AI is widely used in developing countries. For example, in India 34 million inseminations were carried out in 2007 while about 8 million were carried out in Brazil For Africa, Asia and Latin America and the Caribbean regions, AI is mostly used for cattle production (dairy). Other species for which AI is used in all three continents are sheep, goats, horses and pigs. In addition, in Asia, AI is used for chickens, camels, buffaloes and ducks, and in Latin America and Caribbean regions for rabbits, buffaloes, donkeys, alpacas and turkeys. For the most part, semen from exotic breeds is used in local livestock populations. To a lesser extent, semen from local breeds is also used for this purpose. Most of the AI services are provided by the public sector but the contribution of the private sector, breeding organizations and NGOs is also substantial. In Africa and Asia, AI use is concentrated in peri-urban areas. Progesterone monitoring and oestrous synchronization have been applied in a number of developing countries. Applications of oestrous synchronization have been limited to some intensively managed farms where AI is routinely used.

CONCLUSION

The genetics and genomics revolution has at its core information and techniques that can be used to change humanness itself as well as the concepts of what it means to be human. The age-old human fantasies of the mythical chimeras of the ancients, supernatural intelligence, wiping disease from human inheritance, designing a better human being, the fountain of youth, and even immortality now have biotechnical credence in the theoretical promises of genetics and genetic engineering. Not only can humanity's collective genetic inheritance be shaped by selecting which embryos are allowed to develop via pre-implantation genetic diagnosis, but genetic engineering, the availability of the human embryo for experimentation, and combining genes from many species require only sufficient imagination to catalyze the designing of a new humanity. Farm animal genetic resources face a double challenge. On the one hand, the demand for animal products is increasing in developing countries: FAO has estimated that demand for meat will double by 2030; over the same thirty-year period demand for milk will more than double. On the other hand, animal genetic resources are disappearing rapidly worldwide. Over the past 15 years, 300 out of 6 000 breeds identified by FAO have become extinct. Many breeds of local importance for food security are not being improved or utilized in a sustainable manner and are in danger of being lost or diluted by crossbreeding.

Conservation and development of local breeds is important because many of them utilize lower quality feed, are more resilient to climatic stress and to local parasites and diseases, and represent a unique source of genes for improving health and performance traits of industrial breeds. It is also important to develop and utilize local breeds that are already adapted to their environments, most of which are harsh, with very limited natural and managerial input. Animals genetically adapted to these conditions are expected to be more productive at lower costs, support food, agriculture and cultural diversity, and be effective in achieving local food security objectives.

Local communities depend on these adapted genetic resources in many countries. Their disappearance or drastic modification, for example, by crossbreeding, absorption or replacement by exotic breeds, will have serious negative impacts on these human populations.

Presently, most breeds at risk of extinction are not supported by any established conservation programmes or active conservation through sustainable utilization (breeding plans) and therefore breed extinction rates are increasing globally.

THE GLOBAL STRATEGY FOR THE MANAGEMENT OF FARM ANIMAL GENETIC RESOURCES

The key component of the *global strategy* is the country-based planning and implementation infrastructure, which includes five structural elements:

- The *global focal point* at FAO headquarters leads the planning, development and implementation of the overall strategy; develops and maintains the information and communication systems; oversees preparation of guidelines; coordinates regional activity; prepares reports and documents for meetings; facilitates policy discussions; identifies training, education and technology transfer needs; develops programme and project proposals; and mobilizes donor resources.
- *Regional focal points* facilitate regional communications; provide technical assistance and leadership; coordinate regional training, research and planning activities; help develop regional policies; assist in identifying project priorities and proposals, and interact with government agencies, donors, research institutions and non-governmental organizations.
- *National focal points* lead, facilitate and coordinate country activities; identify capacity-building needs; develop project proposals; assist with the development and implementation of country policies; and interface with national stakeholders, regional focal points and the global focal point.
- *Donor and stakeholder involvement* is necessary to provide financial and institutional support to the *global strategy*. In this context, the

global focal point seeks to ensure stakeholder involvement in all major aspects of the *global strategy,* facilitating opportunities for governmental and non-governmental contributions.

- *DAD-IS, the Domestic Animal Diversity Information System,* is a widely available and easily accessible global database and information source. This global facility makes it possible to share data and information among countries, allowing a rapid and cost-effective distribution of guidelines, reports and meeting documents, and provides a platform to exchange views and address specific information requests, linking breeders, scientists and policy-makers. A key feature of DAD-IS is the breeds database, which provides the data for the early warning system for animal genetic resources through the World Watch List for Domestic Animal Diversity, whose third edition was released in 2001.

FIRST REPORT ON THE STATE OF THE WORLD'S ANIMAL GENETIC RESOURCES

As part of the *global strategy for the management of farm animal genetic resources,* FAO invited 188 countries to participate in the first *Report on the State of the World's Animal Genetic Resources,* which is to be completed by 2006. To date, 151 countries have accepted to submit country reports. Guidelines for preparation of country reports have been published in Animal Genetic Resources Information Bulletin (FAO) no. 30. These guidelines are used to assist countries in preparing reports as strategic policy documentation covering the state of animal genetic resources, the state of the art and national capacity to manage these resources, and country needs and priorities. Country reports will serve as the base documentation for the State of the World Reporting Process; thus the involvement of all stakeholders in the development of these reports is strongly encouraged.

The objective of the country and global assessments is to provide a comprehensive analysis of the status and trends of the world's farm animal biodiversity and of their underlying causes, as well as of local knowledge regarding its management.

The task is to go beyond description of the resources by analysing the state of these resources and the capacities to manage them, drawing lessons from past experiences and identifying problems and priorities. Country reports are policy documents covering three strategic questions: *Where are we? Where do we need to be? How do we get to where we need to be?* Country reports are intended to be used in planning and implementing priority country actions. In addition, the country report will serve as documentation for the development of the regional and global reports on strategic priorities for action and, subsequently, the first *Global Report on the State of Farm Animal Genetic Resources.*

Country reports provide an assessment in three major areas:

- The *state of diversity* to evaluate the state of conservation, erosion and utilization of farm animal agricultural biodiversity, and an analysis of the underlying processes;
- The *state of national capacity* to manage animal genetic resources, including existing policies, management plans, institutional infrastructures, human resources and equipment;
- The *state of the art* of the available methodologies and technologies to assist farmers, breeders and scientists to better understand, use, develop and conserve animal genetic resources and thereby contribute to global food security and rural development.

International organizations are also being invited to contribute to the state of the world's animal genetic resources preparatory process by providing reports. The long-term aim of the process is for countries and regions to build on the analyses contained in the country reports in order to plan and implement appropriate management of their farm animal genetic resources.

The first *Report on the State of the World's Animal Genetic Resources* will contain the *Report on Strategic Priorities for Action* and will be based on a synthesis of country reports, thematic studies and reports from international organizations.

FIELDWORK AT COUNTRY AND REGIONAL LEVELS

Countries were requested to nominate a national focal point and designate a national coordinator to facilitate the development of the country network on the management of animal genetic resources and to serve as official contact with the global focal point. Keeping in mind that the process involves both scientific and policy matters, the establishment of a National Consultative Committee is recommended to identify the primary areas and issues that need to be addressed in the preparation of the country report and to oversee its preparation. It is essential that the National Consultative Committee have wide and diverse representation and develop a broad network to ensure opportunities for all stakeholders to contribute to the country report.

The response of countries to the invitation of FAO's Director-General to participate in the first *Report on the State of the World's Animal Genetic Resources* and submit a country report has been very positive. During part of 2001 and 2002, FAO trained almost 400 professionals from 178 countries in the preparation of national reports. At the moment, FAO has a team of 15 consultants working in 14 country groupings in all regions of the world. Most countries have undertaken the organization of national stakeholder workshops to elaborate their animal genetic resources policies leading to the country reports. FAO has organized 14 subregional workshops to discuss draft country reports and regional priorities for action. This has promoted regional cooperation and allows countries that may be experiencing delays to catch up with those in a more advanced state of country report preparation and learn from their experiences. These sessions were coordinated by the regional facilitators acting as FAO consultants.

FAO has provided technical and financial support to 115 countries with contributions from the Governments of the Netherlands and Finland and from the Nordic Gene Bank. FAO and the World Association for Animal Production (WAAP) signed an agreement to provide technical and operational support for the state of the world animal genetic resources reporting process, including training and country follow-up. FAO considers this cooperation a prime example of effective collaboration with an international non-governmental organization.

7

Genetic Modification

GENETICALLY MODIFIED CROPS

In October of 2000, the detection of the genetically engineered StarLink corn in Tacos and several other food products caught the attention of American consumers. During the last week of February, it was reported that most of the corn seeds ready for planting this year are contaminated with StarLink corn. Farmers and federal regulators, consequently, face a dilemma about what corn to plant. Americans are also increasingly aware of protests around the world objecting to genetically modified foods, identifying them as frankenfood or as environmentally unfriendly. These recent news items have re-ignited public debate over the technology of genetically altering crops. This and the subsequent two articles in future issues of this newsletter will attempt to address some common issues related to Genetically Modified Foods and the technology of genetic engineering. Part I will delve into the background of genetic engineering and the future of genetically altered foods; Part II will address issues related to the benefits and drawbacks of this technology; and Part III will delineate, in simple terms, the basics of how genetically modified organisms are created.

Selective plant breeding is not a new concept. Casual selection of observed desirable traits by our ancestors essentially tamed wild plants and made them suitable for agriculture. In the past, if pests devastated a field of crops and a few plants stayed alive and healthy, the seeds from these healthy plants were used to generate the next crop. Thus the beneficial factors that made the plants resistant were transferred to the next generation, making the new generation of crops slightly more resistant to the same pests. Such selections have been used for over 10,000 years, since the beginning of agriculture and have resulted in significant advances for humanity with increased yields, disease resistance and, overall, greater productivity. A good example is corn; the original crop was Teosinte with very small seeds and very few seeds in each cob. Over the centuries people selected for various traits thus improving the size of the corn kernel, the number of kernels in each cob, the stronger attachment of the kernels to the cob so that it could be harvested and so on, resulting in the corn that we are familiar with today. It has since become evident that all traits, beneficial or otherwise, are conferred to living organisms by genes. Genes are small segments of the DNA (deoxyribonucleic acid) that code for specific proteins, which in turn regulate the various traits in all living organisms. There are thousands of genes in plants that regulate the activities governing the growth, stature, colour, and all other aspects of plant growth and survival. Selection through traditional breeding involved the transfer of numerous genes from one generation to the next, including the genes for the beneficial trait as well as the undesirable traits. It then took years of self-breeding and selecting to get a plant with both the normal characteristics as well as the beneficial trait desired.

The advent of genetic engineering greatly enhanced this process of transferring a beneficial trait into plants by directly transferring the gene/s responsible for the beneficial attribute. So in one generation, or one planting season, a plant can be created that is the same in all respects except the addition of the beneficial trait. Since genes in all living organisms code for similar proteins and properties, it is possible to transfer a gene from, say, one good corn variety to another corn variety or from fish to strawberry plants.

We can now define a genetically modified organism: It is any organism that has been modified by altering one or more genes by recombinant techniques. A recombinant technique is the method used to transfer a gene of interest from one organism to another. A genetically modified food therefore is any food that is produced from plants or animals that have been genetically altered using this method. The first genetically altered plant created was a tobacco plant with resistance to antibiotics in 1983. It was almost 10 years later when the first commercial genetically altered crop, a delayed ripening tomato, "Flavr-savr", was commercially released. This was not a commercial success, however, for reasons related to production and marketing strategies. This was soon followed by the release of several crops including Roundup Ready soy and corn. Corn and soy are the two most commonly used food crops that have been genetically altered. They have been primarily altered so that the plants can resist pests, diseases or chemicals used to destroy weeds in the field.

Such alterations that improve the health of the plant and potentially benefit farmers are commonly termed as input traits; that is farmers use fewer inputs to grow their crops, be it pesticide, herbicide or chemicals to prevent diseases. There are other alterations that are possible that alter the property of the oil or starch in the seeds. These are termed output traits; that is the seeds produced by the plant have altered properties either by way of improved yields, nutritional content or higher levels or quality of starch, proteins or oils.

Genetically altered foods are very prevalent, at least in the United States and the Western world. More than 60per cent of the foods we purchase from the supermarket today have ingredients derived from genetically modified crops. Most of these are either from corn or soybeans, which are the base for numerous ingredients manufactured for the food industry, including starch, oils, proteins and other ingredients.

Despite this prevalence, a recent USDA consumer focus group survey revealed that most consumers were unaware of the use of biotechnology in foods. Furthermore, the benefits of biotechnology were viewed as skewed towards producers and manufacturers, with little benefit to the consumer. There was also skepticism related to the long-term health effects and impact on environment. It is therefore essential that we disseminate the information about the technology of genetic engineering so that we can have an informed debate on the merits and shortcomings of this technology.

GENETICALLY MODIFIED ORGANISM AND GENETICALLY ENGINEERED ORGANISM

A genetically modified organism (GMO) or genetically engineered organism (GEO) is an organism whose genetic material has been altered using genetic engineering techniques. These techniques, generally known as recombinant DNA technology, use DNA molecules from different sources, which are combined into one molecule to create a new set of genes. This DNA is then transferred into an organism, giving it modified or novel genes. Transgenic organisms, a subset of GMOs, are organisms which have inserted DNA that originated in a different species. Some GMOs contain no DNA from other species and are therefore not transgenic but cisgenic.

PRODUCTION

Genetic modification involves the insertion or deletion of genes. When genes are inserted, they usually come from a different species, which is a form of horizontal gene transfer. In nature this can occur when exogenous DNA penetrates the cell membrane for any reason. To do this artificially may require attaching the genes to a virus or just physically inserting the extra DNA into the nucleus of the intended host with a very small syringe, or with very small particles fired from a gene gun. However, other methods exploit natural forms of gene transfer, such as the ability of *Agrobacterium* to transfer genetic material to plants, or the ability of lentiviruses to transfer genes to animal cells.

History

The general principle of producing a GMO is to add new genetic material into an organism's genome. This is called genetic engineering and was made possible through the discovery of DNA and the creation of the first recombinant bacteria in 1973, *i.e.*, *E.coli* expressing a Salmonella gene. This led to concerns in the scientific community about potential risks from genetic engineering, which were thoroughly discussed at the Asilomar Conference. One of the main recommendations from this meeting was that government oversight of recombinant DNA research should be established until the technology was deemed safe. Herbert Boyer then founded the first company to use recombinant DNA technology, Genentech, and in 1978 the company announced creation of an *E. coli* strain producing the human protein insulin.

In 1986, field tests of bacteria genetically engineered to protect plants from frost damage (ice-minus bacteria) at a small biotechnology company called Advanced Genetic Sciences of Oakland, California, were repeatedly delayed by opponents of biotechnology. In the same year, a proposed field test of a microbe genetically engineered for a pest resistance protein by Monsanto Company was dropped.

Uses

GMOs have widespread applications. They are used in biological and medical research, production of pharmaceutical drugs, experimental medicine (*e.g.* gene therapy), and agriculture (*e.g.* golden rice). The term "genetically modified organism" does not always imply, but can include, targeted insertions of genes from one species into another. For example, a gene from a jellyfish, encoding a fluorescent protein called GFP, can be physically linked and thus co-expressed with mammalian genes to identify the location of the protein encoded by the GFP-tagged gene in the mammalian cell. Such methods are useful tools for biologists in many areas of research, including those who study the mechanisms of human and other diseases or fundamental biological processes in eukaryotic or prokaryotic cells.

To date the broadest application of GMO technology is patent-protected food crops which are resistant to commercial herbicides or are able to produce pesticidal proteins from within the plant, or *stacked trait* seeds, which do both. The largest share of the GMO crops planted globally are owned by Monsanto Company, according to the company. In 2007, Monsanto's trait technologies were planted on 246 million acres (1,000,000 km^2) throughout the world, a growth of 13 percent from 2006.

In the corn market, Monsanto's triple-stack corn – which combines Roundup Ready 2 weed control technology with YieldGard Corn Borer and YieldGard Rootworm insect control – is the market leader in the United States. U.S. corn farmers planted more than 17 million acres (69,000 km^2) of triple-stack corn in 2007, and it is estimated the product could be planted on 45 million to 50 million acres (200,000 km^2) by 2010. In the cotton market, Bollgard II with Roundup Ready Flex was planted on nearly 3 million acres.

Rapid growth in the total area planted is measurable by Monsanto's growing share. On January 3, 2008, Monsanto Company (MON.N) said its quarterly profit nearly tripled, helped by strength in its corn seed and herbicide businesses, and raised its 2008 forecast. According to the International Service for the Acquisition of Agri-Biotech Applications (ISAAA), of the approximately 8.5 million farmers who grew biotech crops in 2005, some 90per cent were resource-poor farmers in developing countries. These include some 6.4 million farmers in the cotton-growing areas of China, an estimated 1 million small farmers in India, subsistence farmers in the Makhathini flats in KwaZulu Natal province in South Africa, more than 50,000 in the Philippines and in seven other developing countries where biotech crops were planted in 2005.. ISAAA estimated that by 2008, 13.3 million farmers were growing GM crops, including 12.3 million in developing counties,. These comprised 7.1 million in China (Bt cotton), 5.0 million in India (Bt cotton), and 200,000 in the Philippines.

"The Global Diffusion of Plant Biotechnology: International Adoption and Research in 2004", a study by Dr. Ford Runge of the University of Minnesota, estimates the global commercial value of biotech crops grown in the 2003–2004 crop year at US$44 billion.

In the United States the United States Department of Agriculture (USDA) reports on the total area of GMO varieties planted. According to National Agricultural Statistics Service, the States published in these tables represent 81-86 percent of all corn planted area, 88-90 percent of all soybean planted area, and 81-93 percent of all upland cotton planted area (depending on the year). USDA does not collect data for global area. Estimates are produced by the International Service for the Acquisition of Agri-biotech Applications (ISAAA) and can be found in the report, Global Status of Commercialized Transgenic Crops: 2007. Transgenic animals are also becoming useful commercially. On 6 February 2009 the U.S. Food and Drug Administration approved the first human biological drug produced from such an animal, a goat. The drug, ATryn, is an anticoagulant which reduces the probability of blood clots during surgery or childbirth. It is extracted from the goat's milk.

Detection

Testing on GMOs in food and feed is routinely done by molecular techniques like DNA microarrays or qPCR. The test can be based on screening elements or event-specific markers for the official GMOs (like Mon810, Bt11, or GT73). The array-based method combines multiplex PCR and array technology to screen samples for different potential GMOs, combining different approaches (screening elements, plant-specific markers, and event-specific markers). The qPCR is used to detect specific GMO events by usage of specific primers for screening elements or event-specific markers.To avoid any kind of false positive or false negative testing outcome, comprehensive controls for every step of the process is mandatory. A CaMV check is important to avoid false positive outcomes based on virus contamination of the sample.

TRANSGENIC MICROBES

Bacteria were the first organisms to be modified in the laboratory, due to their simple genetics. These organisms are now used for several purposes, and are particularly important in producing large amounts of pure human proteins for use in medicine. Genetically modified bacteria are used to produce the protein insulin to treat diabetes. Similar bacteria have been used to produce clotting factors to treat haemophilia, and human growth hormone to treat various forms of dwarfism. These recombinant proteins are safer than the products they replaced, since the older products were purified from cadavers and could transmit diseases. Indeed the human-derived proteins caused many cases of AIDS and hepatitis C in haemophilliacs and Creutzfeldt-Jakob disease from human growth hormone.

For instance, the bacteria which cause tooth decay are called *Streptococcus mutans*. These bacteria consume leftover sugars in the mouth, producing lactic acid that corrodes tooth enamel and ultimately causes cavities. Scientists have recently modified *Streptococcus mutans* to produce no lactic acid. These transgenic bacteria, if properly colonized in a person's mouth, could reduce the formation of cavities. Transgenic microbes have also been used in recent research to kill or hinder tumours, and to fight Crohn's disease. Genetically modified bacteria are also used in some soils to facilitate crop growth, and can also produce chemicals which are toxic to crop pests.

TRANSGENIC ANIMALS

Transgenic animals are used as experimental models to perform phenotypic tests with genes whose function is unknown. Genetic modification can also produce animals that are susceptible to certain compounds or stresses for testing in biomedical research. Other applications include the production of human hormones such as insulin. In biological research, transgenic fruit flies (*Drosophila melanogaster*) are model organisms used to study the effects of genetic changes on development.

Fruit flies are often preferred over other animals due to their short life cycle, low maintenance requirements, and relatively simple genome compared to many vertebrates. Transgenic mice are often used to study cellular and tissue-specific responses to disease.

This is possible since mice can be created with the same mutations that occur in human genetic disorders, the production of the human disease in these mice then allows treatments to be tested. In 2009 scientists in Japan announced that they had successfully transferred a gene into a primate species (marmosets) and produced a stable line of breeding transgenic primates for the first time. It is hoped that this will aid research into human diseases that cannot be studied in mice, for example Huntington's disease and strokes. Cnidarians such as *Hydra* have become attractive model organisms to study the evolution of immunity.

For analytical purposes an important technical breakthrough was the development of a transgenic procedure for generation of stably transgenic hydras by embryo microinjection.

Transgenesis in fish with promoters driving an overproduction of "all fish" growth hormone has resulted in dramatic growth enhancement in several species, including salmonids, carps and tilapias. These fish have been created for use in the aquaculture industry to increase the speed of development and potentially, reduce fishing pressure on wild stocks. None of these GM fish have yet appeared on the market, mainly due to the concern expressed among the public of the fish's potential negative effect on the ecosystem should they escape from fish farms.

Gene Therapy

Gene therapy, uses genetically modified viruses to deliver genes that can cure disease into human cells. Although gene therapy is still relatively new, it has had some successes. It has been used to treat genetic disorders such as severe combined immunodeficiency, and treatments are being developed for a range of other currently incurable diseases, such as cystic fibrosis, sickle cell anemia, and muscular dystrophy.. Current gene therapy technology only targets the Non-reproductive cells meaning that any changes introduced by the treatment can not be transmitted to the next generation. Gene therapy targeting the reproductive cells-so called "Germ line Gene Therapy"-is very controversial and is unlikely to be developed in the near future.

Transgenic Plants

Transgenic plants have been engineered to possess several desirable traits, including resistance to pests, herbicides or harsh environmental conditions, improved product shelflife, and increased nutritional value. Since the first commercial cultivation of genetically modified plants in 1996, they have been modified to be tolerant to the herbicides glufosinate and glyphosate, to be resistant to virus damage as in Ringspot virus resistant GM papaya, grown in Hawaii, and to produce the Bt toxin, a potent insecticide.

Bt-maize is a corn that has been genetically modified by splicing the toxin-producing gene from bacteria into the DNA sequence of the corn in order to sicken or kill insects that try to consume it. While some genetically modified crops are more nutritious because they contain these extra vitamins and minerals, they will not cure all of the malnutrition-related ailments in the world and should only be a supplement to a balanced diet.

Cisgenic Plants

Genetically modified sweet potatoes have been enhanced with protein and other nutrients, while golden rice, developed by the International Rice Research

Institute, has been discussed as a possible cure for Vitamin A deficiency. In reality, customers would have to eat twelve bowls of rice a day in order to meet the recommended levels of Vitamin A. In January 2008, scientists altered a carrot so that it would produce calcium and become a possible cure for osteoporosis; however, people would need to eat 1.5 kilograms of carrots per day to reach the required amount of calcium.

The coexistence of GM plants with conventional and organic crops has raised significant concern in many European countries. Since there is separate legislation for GM crops and a high demand from consumers for the freedom of choice between GM and non-GM foods, measures are required to separate foods and feed produced from GMO plants from conventional and organic foods. European research programmes such as Co-Extra, Transcontainer and SIGMEA are investigating appropriate tools and rules. At the field level, biological containment methods include isolation distances and pollen barriers.

CONTROVERSY

Biological Process

The use of GMOs has sparked significant controversy in many areas. Some groups or individuals see the generation and use of GMO as intolerable meddling with biological states or processes that have naturally evolved over long periods of time, while others are concerned about the limitations of modern science to fully comprehend all of the potential negative ramifications of genetic manipulation.

Foodchain

The safety of GMOs in the foodchain has been questioned, with concerns such as the possibilities that GMOs could introduce new allergens into foods, or contribute to the spread of antibiotic resistance.

Although scientists have assured consumers of the safety of these types of crops, consumption has been discouraged in many countries by food and environmental activist groups who protest GM crops, claiming they are unnatural and therefore unsafe. This has led to the adoption of laws and regulations that require safety testing of any new organism produced for human consumption.

Trade with Europe and Africa

In response to negative public opinion, Monsanto announced its decision to remove their seed cereal business from Europe, and environmentalists crashed a World Trade Organization conference in Cancun that promoted GM foods and was sponsored by Committee for a Constructive Tomorrow (CFACT). Some African nations have refused emergency food aid from developed countries, fearing that the food is unsafe. During a conference in the Ethiopian capital of Addis Ababa, Kingsley Amoako, Executive Secretary of the United Nations Economic Commission for Africa (UNECA), encouraged African nations to accept genetically modified food and expressed dissatisfaction in the public's negative opinion of biotechnology.

TYPES OF TOXICOLOGICAL HAZARDS TO CONSUMERS AND PRODUCERS ASSOCIATED WITH BD FOODS

Current techniques of developing organisms used in the production of BD foods typically involve the transfer to the host of the desired gene or genes in combination with a promoter and a gene for a selectable marker trait that allows the efficient isolation of cells or organisms that have been transformed from those that have not. Common selectable markers in plants have included resistance to antibiotics (kanamycin/neomycin or ampicillin) or herbicides.

Several key issues have been raised with respect to the potential toxicity associated with BD foods, including the inherent toxicity of the transgenes and their products, and unintended (pleiotropic or mutagenic) effects resulting from the insertion of the new genetic material into the host genome. Unintended effects of gene insertion might include an over-expression by the host of inherently toxic or pharmacologically active substances, silencing of normal host genes, or alterations in host metabolic pathways. It is important to recognize that, with the exception of the introduction of marker genes, the process of genetic engineering does not, in itself, create new types of risk.

FUTURE CHALLENGES IN THE ASSESSMENT OF THE SAFETY OF BD FOODS

Current safety assessment methodologies are focused primarily on the evaluation of the toxicity of single chemicals. Food is a complex mixture of many chemicals. Using animal models, the evaluation of most aspects of the safety of single components of the diet, such as a Bt toxin, is possible using widely accepted protocols. Future projects may involve more complicated manipulations of plant chemistry. In this case, safety testing will be more challenging. Whole foods cannot be tested with the high dose strategy currently used for single chemicals to increase the sensitivity in detecting toxic endpoints. Also, the question of potential deleterious interactions between new or enhanced levels of known toxic agents in BD foods will undoubtedly be raised. The safety testing of multiple combinations of chemicals remains a difficult proposition for toxicologists. In view of these challenges, there is a clear need for the development of effective protocols to allow the assessment of the safety of whole foods. The responsibility of toxicologists is to assess whether foods derived through biotechnology are at least as safe as their conventional counterparts and to ascertain that any levels of additional risk are clearly defined. In achieving this goal, it is important to recognize that it is the food product itself, rather than the process through which it is made, that should be the focus of attention. In assessing safety, the use of the substantial equivalency concept provides guidance as to the nature of any new hazards.

Scientific analysis indicates that the process of BD food production is unlikely to lead to hazards of a different nature from those already familiar to toxicologists. The safety of current BD foods, compared with their conventional counterparts, can be assessed with reasonable certainty using established and accepted methods of analytical, nutritional, and toxicological research.

A significant limitation may occur in the future if transgenic technology results in more substantial and complex changes in a foodstuff. Methods have not yet been developed by which whole foods (as compared with single chemical components) can be fully evaluated for safety. Progress also needs to be made in developing definitive methods for the identification and characterization of protein allergens, and this is currently a major focus of research. Improved methods of profiling plant and microbial metabolities, proteins, and gene expression may be helpful in detecting unexpected changes in BD organisms and in establishing substantial equivalence. The level of safety of current BD foods to consumers appears to be equivalent to that of traditional foods. Verified records of adverse health effects are absent, although the current passive reporting system would probably not detect minor or rare adverse effects, nor can it detect a moderate increase in common effects such as diarrhea. However, this is no guarantee that all future genetic modifications will have such apparently benign and predictable results. A continuing evolution of toxicological methodologies and regulatory strategies will be necessary to ensure that this level of safety is maintained.

GENETIC MODIFICATION IN AGRICULTURE

Genetic modification, otherwise referred to as recombinant DNA (rDNA) technology or gene-splicing, has proven to be a more precise, predictable and better understood method for the manipulation of genetic material than previously attained through conventional plant breeding. To date, agricultural applications of the technology have involved the insertion of genes for desirable agronomic traits (*e.g.* herbicide tolerance, insect resistance) into a variety of crop plants, and from a variety of biological sources. Examples include soybeans modified with gene sequences from a *Streptomyces* species encoding enzymes that confer herbicide tolerance, and corn plants modified to express the insecticidal protein of an indigenous soil microorganism, *Bacillus thuringiensis* (Bt). A growing body of evidence suggests that the technology may be used to make enhancements to not only the agronomic properties, but the food, nutritional, industrial and medicinal attributes of genetically modified (GM) crops. Regulatory supervision of rDNA technology and its products has been in place for a longer period of time in the United States than in most other parts of the world. The methods and approaches established to evaluate the safety of products developed using rDNA technology continue to evolve in response to the increasing availability of new scientific information. As our understanding of the potential applications of the technology is broadened, the safety of products developed using rDNA technology and the potential effects of introduced gene sequences on human health or the environment will be more closely scrutinized. In fact, much of the knowledge acquired during the commercialization of the products of rDNA technology in agriculture is now finding application in evaluating the safety of products developed through more conventional means.

The objective of this chapter is to provide the reader with an overview of the significant events leading up to the present, science-based, regulatory framework that exists for the safety evaluation of GM food crops within the United States. An attempt has been made to discuss concerns over the sufficiency of existing regulations, as well as to highlight recent initiatives taken by federal regulatory agencies to address them. Through better communication of how the regulatory process functions within the United States, it is anticipated that current and future applications of rDNA technology in agriculture will be met with a greater level of understanding and acceptance.

HISTORICAL PERSPECTIVE

rDNA technology was first developed in the 1970s. The initial response of the scientific community, including members of the National Academy of Science (NAS), to the prospects of rDNA technology, was to postpone any further research involving the technology until the potential risks to human health and the environment could be evaluated.

Researchers attending the International Conference on Recombinant DNA Molecules in 1975, otherwise known as the Asilomar Conference, tried to establish a scientific consensus on how best to self-regulate emerging applications of the technology. The conditions and restrictions that were proposed at this conference have formed the basis by which federal guidelines and policies for rDNA technology research were drafted within the United States.

National Institutes of Health (NIH)

The National Institutes of Health (NIH) was the first federal regulatory agency to publish their interests in evaluating the safety of rDNA technology in 1976, in the form of guidelines for the conduct of research. Because of the uncertainties that existed at the time, all research into the potential applications of rDNA technology was limited to the confines of federally funded labouratories under NIH control. After continued research, and a more careful assessment and monitoring of the risks, a set of less restrictive guidelines was published in 1978. However, the environmental release of organisms developed using rDNA technology outside the confines of controlled labouratory conditions was prohibited unless otherwise approved by the NIH director. In the early 1980s, the NIH established an rDNA Advisory Committee (RAC) to review all data and experience gained with applications of the technology under its control. Based on recommendations of the RAC, a more relaxed set of research guidelines was published by the NIH in 1983.

The NIH approved the first environmental release of an organism developed using rDNA technology (ice-minus strain of *Pseudomonas*) in 1983. In response, they were criticized for failing to prepare a statement or assessment of the environmental impact of their regulatory decision as required under the National Environmental Policy Act (NEPA). Once the legal controversy had subsided, all responsibility that the NIH had for regulating the environmental introduction of GM organisms was relinquished. Nevertheless, NIH guidelines continue to be referenced in assessing the safety of rDNA research performed within industry, federal and other state labouratories. However, it was unclear which federal regulatory agencies would be responsible for ensuring the safety of the products developed using rDNA technology.

Office of Science and Technology Policy (OSTP)

In response to a need for clarification, the Office of Science and Technology Policy (OSTP) began work on the development of a policy to establish a federal regulatory framework for evaluating the safety of products developed using rDNA technology. According to the Coordinated Framework, the products of rDNA technology should be regulated on the basis of the unique characteristics and features that they exhibit, not their method of production. The products of rDNA technology were considered to pose risks to human health and the environment similar to those posed by conventional products already regulated within the United States. As a result, no new federal regulatory agencies or regulations were required. The Coordinated Framework did not, however, rule out the possibility of the development of new guidelines, procedures, criteria or even regulations to supplement or alter the scope of existing statutes for the products of rDNA technology.

The Coordinated Framework identified three federal regulatory agencies within the United States: the US Food and Drug Administration (US FDA), the US Department of Agriculture (USDA) and the US Environmental Protection Agency (US EPA), as having primary responsibilities for evaluating the products of rDNA technology under development at that time.

In 1992, the OSTP released another document entitled, 'Exercise of Federal Oversight within the Scope of Statutory Authority: Planned Introductions of Biotechnology Products Into the Environment', outlining the proper basis by which federal regulatory agencies were expected to exercise their regulatory authority. As with conventional products, dependent upon the intended use and function, more than one federal regulatory agency may share an interest in evaluating the safety of a product developed using rDNA technology. If more than one federal regulatory agency has an interest, lead agencies are identified as being responsible for coordinating activities to limit any potential duplication of efforts. Although federal regulatory agencies worked independent of one another, it was realised that close working relationships would need to be established in order to evaluate effectively the safety of products developed using rDNA technology.

Recently, the OSTP teamed up with the White House Council on Environmental Quality (CEQ) to perform a six-month inter-agency evaluation of the federal regulatory agency responsibilities in evaluating the environmental safety of products developed using rDNA technology. A case-study approach for a variety of different classes of products developed using rDNA technology was used to evaluate the level of federal regulatory agency involvement, to identify strengths, weaknesses and areas of potential improvement. The review concluded that none of the previously approved products of rDNA technology has had any significant negative impact on the environment.

Although all the case studies were published, OSTP/CEQ failed to reach a consensus on issues relating to the relevant strengths and weaknesses of the existing regulatory structure within the time allotted for the completion of its review. A review of the case studies published provides a comprehensive interpretation of the responsibilities of each federal regulatory agency in ensuring the safety of the products developed using rDNA technology.

National Academy of Sciences (NAS)

The National Academy of Sciences (NAS), and its operating arm, the National Research Council (NRC), have served as a primary source of scientific, technological, human health and environmental policy advice during the development of regulatory approaches for the safety evaluation of products developed using rDNA technology within the United States.

In 1987, the NRC published a report concerning the potential human health and environmental hazards associated with the commercial introduction of GM organisms, entitled *Introduction of Recombinant DNA-Engineered Organisms into the Environment: Key Issues*. The risks associated with the introduction of GM organisms were considered to be essentially the same in kind as those associated with unmodified organisms.

In other words, rDNA technology did not appear to introduce any unique risks as compared to the products that had been developed using more conventional methods of genetic modification. In reaching these conclusions, the NRC performed an evaluation of the similarities and differences in the properties exhibited by products developed using a variety of different techniques. To this day, the conclusions of this report continue to be referenced by the regulators and developers of GM crop varieties worldwide.

A subsequent NRC report, entitled *Field Testing Genetically Modified Organisms: Framework for Decisions*, also reached similar conclusions, but provided additional guidance as to how regulatory decisions concerning the introduction of GM organisms should be made. The NRC recommended that regulatory decisions concerning the introduction of GM organisms should be made on a case-by-case basis. Consistent with the Coordinated Framework, the NRC did not consider the nature of the process used for the genetic modification of an organism to be a useful criterion for determining whether a product requires less or more regulatory oversight. As a result, no valid reason existed to regulate organisms genetically modified via modern techniques (*e.g.* rDNA technology) any differently from organisms genetically modified via more conventional means. Similar conclusions have been published in the reports of international standards-setting organizations. In retrospect, within both reports, the NRC acknowledged that modern and conventional methods of genetic modification are not without risks to human health or the environment, and as a result, neither could be considered inherently more risky.

As a result, regulatory decisions concerning the safety of the products of rDNA technology need only to take into consideration the specific characteristics exhibited by a particular GM organism and the environment in which it is to be introduced, and not the method by which it has been produced. The NRC has further articulated their conclusions into what is now commonly referred to as the 'Concept of Familiarity'.

Although familiarity with the characteristics of a particular organism or the environment to which it will be introduced would not necessarily mean it was safe, it can be expected to provide a sufficient amount of information to allow for a judgement to be made of the risks. For example, familiarity with a new GM plant variety could be established based on comparisons between characteristics of the parent line or other crop species exhibiting similar traits, as well as through the results of actual field tests involving the GM plant. These principles were further elabourated upon by the Organisation for Economic Co-operation and Development (OECD), and as a result, have been referenced in the development of regulatory policies for evaluating the safety of GM crops on a global basis.

More recently, a committee established by the NRC published a report entitled *Genetically Modified Pest-Protected Plants: Science and Regulation*, based on a review of all scientific and regulatory data collected during the regulatory approval process for GM crops within the United States. The primary objective was to assess independently the effectiveness of existing and proposed regulations for the safety evaluation of GM crops expressing plant pesticides (*i.e.* plant-incorporated protectants).

No new evidence was identified to suggest plants expressing plant pesticides posed any greater risk to human health or the environment as a result of their genetic modification. In fact, the NRC concluded, 'with careful planning and appropriate regulatory oversight, commercial cultivation of GM plants is not expected to pose higher risks and may pose less risk than other commonly used chemical and biological pest-management techniques'.

However, the NRC report included requests for federal regulatory agencies to further strengthen the current regulatory approval process through better coordination and communication between agencies, on-going investment in the research and monitoring of potential human health (*e.g.* allergenicity) and environmental impacts (*e.g.* insect resistance), and by providing greater access to information evaluated in support of regulatory decisions.

PRODUCTS OF AGRICULTURAL BIOTECHNOLOGY

The regulatory approach to the safety evaluation of plants developed using rDNA technology has evolved in the best interests of research scientists, industry and the general public. The agricultural products of rDNA technology, such as GM foods and crops, may require approvals from up to three regulatory agencies; the US FDA, the USDA and the US EPA, depending upon the characteristics exhibited by the GM plant, its proposed use and introduced traits. The same standards of safety are applied to all products regardless of the technology used in their development.

The US FDA is responsible for ensuring the human safety of all new foods and food components, including products developed using rDNA technology, under the Federal Food, Drug, and Cosmetic Act (FFDCA). The USDA evaluates the potential of a GM plant to become a plant pest following its environmental introduction under the Federal Plant Pest Act (FPPA). The US EPA evaluates pesticides, including plant systems modified to express pesticides (*e.g.* insect-protected or virus resistance), under the Federal Insecticide, Fungicide, and Rodenticide Act (FIFRA). As a result, the expression of an insecticidal protein in a food crop would undergo review by the USDA, US EPA and US FDA; a GM food crop exhibiting a modified oil content would be evaluated by the USDA and US FDA; and a non-food horticultural plant developed using rDNA technology for any other purpose (*e.g.* flower colour) would be subject to review by the USDA alone.

In most instances, obtaining all necessary approvals for the commercialization of an agricultural crop developed using rDNA technology takes a decade or more. However, the exact amount of time required will depend on the need to confirm performance, to evaluate characteristics of the food, environmental effects, and to produce the required amount of seed before the product can be distributed and commercially grown by farmers. Up to five years of field trials (5-10 generations of plants) are required for the developer of a new plant variety to collect sufficient data to meet the reporting requirements of the USDA. An additional five months to two years may be required for the US FDA, USDA and/or US EPA to complete all necessary product consultations, reviews and approvals.

Approval for the first commercial planting of a GM food crop was not issued until 1995. Since 1995, more than 40 new agricultural crops developed using rDNA technology have received approval for commercial planting within the United States. In 1999, approximately 72per cent of the total 39.9 million hectares (more than 98 million acres) of GM crops grown worldwide were planted in the United States. Herbicide-tolerant soybeans (54per cent), Bt corn (19per cent) and herbicide tolerant canola (9per cent) accounted for approximately 82per cent of the GM plants cultivated. With an increasing number of agricultural biotechnology products reaching later stages of commercial development, it is anticipated that the overall area planted with GM crops will continue to rise. The regulatory approach to evaluating the human health and environmental safety of GM crops within the United States is best described as a science-based, case-by-case assessment of hazards and risks. This approach has provided the flexibility required to reduce the regulatory burden placed on products that have been determined to be of low risk or concern. All agencies involved in the regulation of plants developed using rDNA continue to implement and develop policies based on recommendations made within the Coordinated Framework. In the future, the US FDA, USDA and the US EPA will be dedicating additional resources towards communicating how GM food and food components are regulated within the United States, and how these regulations function to be protective of both human health and the environment.

FOOD AND FOOD COMPONENTS

The US FDA is responsible for ensuring the safety and wholesomeness of all food and food components, including the products of rDNA technology, under the FFDCA. The US FDA has the authority for the immediate removal of any product from the market that poses potential risk to public health or that is being sold without all necessary regulatory approvals. As a result, a legal burden is placed on developers and food manufacturers to ensure the commodities utilized and foods available to consumers are safe and in compliance with all legal requirements of the FFDCA. In order to understand the regulatory approach followed by the US FDA in the safety evaluation of GM crops, it is useful to consider food and food safety from a historical context.

People had been consuming foods derived from agricultural crops for many years prior to the existence of any food laws or regulations within the United States. Based on this experience, agricultural crops have been accepted as being safe for consumption as food, without additional testing to demonstrate their safety. As long as the new crop variety has exhibited similar agronomic properties, and an appropriate taste and appearance, it has been considered safe to consume. As a result, most foods consumed today, in particular whole foods (*i.e.* fruits, grains and vegetables) and conventional foods, have not been subject to any kind of premarket review or approval by the US FDA. Nevertheless, food scientists have a good understanding that many of the commonly consumed agricultural crops contain natural toxicants (*e.g.* tomatine in tomatoes, solanine in potatoes, cucurbiticin in cucumber, psoralens in celery, etc.). As a result, new plant varieties may be subject to routine chemical analyses to ensure that none of these substances is present at potentially harmful levels. This type of general approach has been used in assessing the safety of thousands of new plant varieties that have been developed over a number of decades of crop breeding without compromising the safety of whole foods.

Role

Consistent with recommendations within the Coordinated Framework, the US FDA considered existing provisions of the FFDCA to be sufficient for the regulation of foods and food components developed using rDNA technology. It was concluded that the scientific and regulatory issues posed by the products of rDNA technology were not significantly different from those posed by conventional products. As a result, GM foods and food components have been subject to the same standards of safety as already exist for the regulation of other foods and food components under the FFDCA. In order to better communicate interpretations of existing provisions of the FFDCA as they relate to the safety evaluation of foods derived from new plant varieties, including the products of rDNA technology, the US FDA released a policy statement in 1992 entitled 'Statement of policy: foods derived from new plant varieties'.

The US FDA has considered the use of genetic modification (*i.e.* rDNA technology) in the development of new plant varieties to represent a continuum of conventional plant breeding practices (*e.g.* mutagenesis, hybridization, protoplast fusion, etc.), and as a result, the safety evaluation of all new plant varieties, not just those developed using rDNA technology, have been evaluated based on an objective analysis of the characteristics of a food or its components, and not on its method of production.

Definition and Scope of Bioengineered Foods

The recently proposed rule of the US FDA concerns 'bioengineered foods' which have been defined as 'foods derived from plant varieties that are developed using *in vitro* manipulations of DNA (generally referred to as rDNA technology)'. As a result, the proposed rule has a much narrower focus than the 1992 US FDA Statement of Policy. The US FDA has explained the need for a change in emphasis based on their expectations that many of the new plant varieties exhibit a greater potential to 'contain substances that are significantly different from, or that are present in food at a significantly different level than before'. As a result, the substances present in foods and food components derived from new plants developed using rDNA technology are less likely to be considered GRAS, and as a result, will require pre-market approval from the US FDA.

Safety and Nutritional Evaluation

The primary objective of the safety and nutritional evaluation is to demonstrate that the food derived from a new plant variety is as safe or nutritious as foods already consumed as a part of the diet. For new plant varieties, including those developed using rDNA technology, a science-based approach is used to focus the evaluation on the demonstrated characteristics of the food or food component. The evaluation of a GM food or food component typically involves reviewing information or data on any newly introduced substances, the known levels of toxicants, as well as the nutritional composition of the plant following modification. Substances that raise safety concerns (*e.g.* toxicants, allergens) would be subject to more extensive evaluation, since both intended and unintended changes may affect the levels of toxicants and nutrients in a food following the modification.

Guidance for performing a safety and nutritional evaluation was provided in the 1992 US FDA Statement of Policy, in a series of flow charts and text that cover:

- The crop that has been modified;
- Source(s) of the introduced genetic material;
- New substances intentionally added to the food as a result of the genetic modification (*e.g.* proteins, but also fatty acids, and carbohydrates).

Documentation required to support the evaluation typically includes: the purpose or intended technical effect of the modification on the plant, together with a description of the various applications or uses, a molecular characterization of the modification including the identities, sources and functions of introduced genetic material; information on the expressed protein products encoded by introduced genes; information relating to the known or suspected allergenicity and toxicity of any expressed gene products; for foods known to cause allergy, information on whether the endogenous allergens have been altered by the genetic modification; information on the compositional and nutritional characteristics of the foods, including anti-nutrients; and in some instances, comparative results of feeding studies involving the foods derived from plants modified using rDNA technology and the non-modified counterpart.

In performing its evaluation, the US FDA is particularly interested in the identification of inherent toxicants, known or potential allergens, assessing the concentration and bioavailability of essential nutrients, the safety and nutritional value of any newly introduced proteins, and the identity, composition, and nutritional value of modified carbohydrates, fats and oils.

If additional questions of safety remain following this evaluation, further toxicological studies may need to be performed. It is recognized that absolute assurance of the safety of any food does not exist. As a result, the goal of the safety evaluation is to establish a reasonable certainty of no harm under anticipated conditions of consumption. With this in mind, experience with the existing food supply has provided the basis for evaluating the safety of new food or food components. Both the Food Advisory Committee and Committee for Veterinary Medicine have been extensively involved in the development of approaches for the safety and nutritional evaluation of foods and food components derived from new plant varieties, including those developed using rDNA technology.

AGRICULTURAL SURPLUSES

Patrick Mulvany, Chairman of the UK Food Group, accused some governments, especially the Bush administration, of using GM food aid as a way to dispose of unwanted agricultural surpluses. The UN blamed food companies and accused them of violating human rights, calling on governments to regulate these profit-driven firms. It is true that the acceptance of biotechnology and genetically modified foods will also benefit rich research companies and could possibly benefit them more than consumers in underdeveloped nations.

LABELLING

While some groups advocate the complete prohibition of GMOs, others call for mandatory labelling of genetically modified food or other products. Other controversies include the definition of patent and property pertaining to products of genetic engineering.

UNDERDEVELOPED NATIONS

Some groups believe that underdeveloped nations will not reap the benefits of biotechnology because they do not have easy access to these developments, cannot afford modern agricultural equipment, and certain aspects of the system revolving around intellectual property rights are unfair to undeveloped countries. For example, The CGIAR (Consultative Group of International Agricultural Research) is an aid and research organization that has been working to achieve sustainable food security and decrease poverty in undeveloped countries since its formation in 1971.

In an evaluation of CGIAR, the World Bank praised its efforts but suggested a shift to genetics research and productivity enhancement. This plan has several obstacles such as patents, commercial licenses, and the difficulty that third world countries have in accessing the international collection of genetic resources and other intellectual property rights that would educate them about modern technology. The International Treaty on Plant Genetic Resources for Food and Agriculture has attempted to remedy this problem, but results have been inconsistent. As a result, "orphan crops", such as tef, millets, cowpeas, and indigenous plants, are important in the countries where they are grown, but receive little investment.

PRIVATE INVESTMENTS

The development and implementation of policies designed to encourage private investments in research and marketing biotechnology that will meet the needs of poverty-stricken nations, increased research on other problems faced by poor nations, and joint efforts by the public and private sectors to ensure the efficient use of technology developed by industrialized nations have been suggested. In addition, industrialized nations have not tested GM technology on tropical plants, focusing on those that grow in temperate climates, even though undeveloped nations and the people that need the extra food live primarily in tropical climates. Many European scientists are disturbed by the fact that political factors and ideology prevent unbiased assessment of the GM technology in some EU countries, with a negative effect on the whole community.

TRANSGENIC ORGANISMS

Another important controversy is the possibility of unforeseen local and global effects as a result of transgenic organisms proliferating. The basic ethical issues involved in genetic research are discussed in the article on genetic engineering.

Some critics have raised the concern that conventionally-bred crop plants can be cross-pollinated (bred) from the pollen of modified plants. Pollen can be dispersed over large areas by wind, animals and insects. In 2007, the U.S. Department of Agriculture fined Scotts Miracle-Gro $500,000 when modified genetic material from creeping bentgrass, a new golf-course grass Scotts had been testing, was found within close relatives of the same genus (*Agrostis*) as well as in native grasses up to 21 km (13 miles) away from the test sites, released when freshly cut grass was blown by the wind.

GM proponents point out that outcrossing, as this process is known, is not new. The same thing happens with any new open-pollinated crop variety—newly introduced traits can potentially cross out into neighbouring crop plants of the same species and, in some cases, to closely related wild relatives. Defenders of GM technology point out that each GM crop is assessed on a case-by-case basis to determine if there is any risk associated with the outcrossing of the GM trait into wild plant populations. The fact that a GM plant may outcross with a related wild relative is not, in itself, a risk unless such an occurrence has negative consequences. If, for example, a herbicide resistance trait was to cross into a wild relative of a crop plant it can be predicted that this would not have any consequences except in areas where herbicides are sprayed, such as a farm. In such a setting the farmer can manage this risk by rotating herbicides. The European Union funds research programmes such as Co-Extra, that investigate options and technologies on the coexistence of GM and conventional farming. This also includes research on biological containment strategies and other measures, to prevent outcrossing and enable the implementation of coexistence.

If patented genes are outcrossed, even accidentally, to other commercial fields and a person deliberately selects the outcrossed plants for subsequent planting then the patent holder has the right to control the use of those crops. This was supported in Canadian law in the case of Monsanto Canada Inc. v. Schmeiser.

"TERMINATOR" AND "TRAITOR"

An often cited controversy is a "Technology Protection" technology dubbed 'Terminator'. This yet-to-be-commercialized technology would allow the production of first generation crops that would not generate seeds in the second generation because the plants yield sterile seeds. The patent for this so-called "terminator" gene technology is owned by *Delta and Pine Land Company* and the United States Department of Agriculture. Delta and Pine Land was bought by Monsanto Company in August 2006. Similarly, the hypothetical Trait-specific Genetic Use Restriction Technology, also known as 'Traitor' or 'T-gut', requires application of a chemical to genetically modified crops to reactivate engineered traits. This technology is intended both to limit the spread of genetically engineered plants, and to require farmers to pay yearly to reactivate the genetically engineered traits of their crops. Traitor is under development by companies including Monsanto and AstraZeneca.

In addition to the commercial protection of proprietary technology in self-pollinating crops such as soybean (a generally contentious issue), another purpose of the terminator gene is to prevent the escape of genetically modified traits from cross-pollinating crops into wild-type species by sterilizing any resultant hybrids. The terminator gene technology created a backlash amongst those who felt the technology would prevent reuse of seed by farmers growing such terminator varieties in the developing world and was ostensibly a means to exercise patent claims.

Use of the terminator technology would also prevent "volunteers", or crops that grow from unharvested seed, a major concern that arose during the Starlink debacle. There are technologies evolving which contain the transgene by biological means and still can provide fertile seeds using fertility restorer functions. Such methods are being developed by several EU research programmes, among them Transcontainer and Co-Extra.

PRODUCTION OF TOXINS

Some bacterial toxins are utilized as invasins because they act locally to promote bacterial invasion. Examples are extracellular enzymes that degrade tissue matrices or fibrin, allowing the bacteria to spread. This includes collagenase, hyaluronidase and streptokinase. Other toxins, also considered invasins, degrade membrane components, such as phospholipases and lecithinases. The pore-forming toxins that insert a pore into eucaryotic membranes are considered as invasins.

The level of risk of these gene products to consumers and those involved in food production can be and is evaluated by standard toxicological methods. The toxicology testing for the Bt endotoxins typifies this approach and has been described in detail by the U.S. EPA. The safety of most Bt toxins is assured by their easy digestibility as well as by their lack of intrinsic activity in mammalian systems. In this case, the good understanding of the mechanism of action of Bt toxins, and the selective nature of their biochemical effects on insect systems, increases the degree of certainty of the safety evaluations. However, each new transgenic product must be considered individually, based on exposure levels and its potency in causing any toxic effects, as is typical of current risk assessment paradigms for chemical agents.

BACTERIAL TOXIGENESIS

Toxigenesis, or the ability to produce toxins, is an underlying mechanism by which many bacterial pathogens produce disease. At a chemical level, there are two main types of bacterial toxins, lipopolysaccharides, which are associated with the cell wall of Gram-negative bacteria, and proteins, which are released from bacterial cells and may act at tissue sites removed from the site of bacterial growth. The cell-associated toxins are referred to as endotoxins and the extracellular diffusible toxins are referred to as exotoxins.

Endotoxins are cell-associated substances that are structural components of bacteria. Most endotoxins are located in the cell envelope. In the context of this article, endotoxin refers specifically to the lipopolysaccharide (LPS) or lipooligosaccharide (LOS) located in the outer membrane of Gram-negative bacteria. Although structural components of cells, soluble endotoxins may be released from growing bacteria or from cells that are lysed as a result of effective host defence mechanisms or by the activities of certain antibiotics. Endotoxins generally act in the vicinity of bacterial growth or presence.

Exotoxins are usually secreted by bacteria and act at a site removed from bacterial growth. However, in some cases, exotoxins are only released by lysis of the bacterial cell. Exotoxins are usually proteins, minimally polypeptides, that act enzymatically or through direct action with host cells and stimulate a variety of host responses. Most exotoxins act at tissue sites remote from the original point of bacterial invasion or growth. However, some bacterial exotoxins act at the site of pathogen colonization and may play a role in invasion.

PRODUCTION OF ALLERGENS

Allergenicity is one of the major concerns about food derived from transgenic crops. However, it is important to keep in mind that eating conventional food is not risk-free; allergies occur with many known and even new conventional foods. For example, the kiwi fruit was introduced into the U.S. and the European markets in the 1960s with no known human allergies; however, today there are people allergic to this fruit. The issues that have to be addressed regarding the potential allergenicity of BD foods are:

- Do the products of novel genes have the ability to elicit allergic reactions in individuals who are already sensitized to the same, or a structurally similar, protein?
- Will transgenic techniques alter the level of expression of existing protein allergens in the host crop plant?
- Do the products of novel genes engineered into food plants have the ability to induce *de novo* sensitization among susceptible individuals?

Considerable scientific resources are being committed to determine the most appropriate and accurate approaches for identifying and characterizing potentially allergenic proteins.

The first systematic approach to allergenicity assessment was developed by the International Life Sciences Institute (ILSI) in collaboration with the International Food Biotechnology Council and was published in 1996. The hierarchical approach described therein has been reviewed and revised by the World Health Organization (WHO) and the Food and Agriculture Organization of the United Nations (FAO) (FAO/WHO, 2001b).

The main approaches currently used in the evaluation of allergenicity are:

- *Determinations of structural similarity, sequence homology, and serological identity*. The objective is to determine whether, and to what extent, the novel protein of interest resembles other proteins that are known to cause allergy among human populations. There are essentially three generic approaches. The first is to examine the overall structural similarity between the protein of interest and known allergens. The second is to determine, using appropriate databases, whether the novel protein is similar to known allergens with respect to either overall amino acid homology, or to discrete areas of the molecule where complete sequence identity with a known allergen may indicate the presence of shared epitopes. The third approach is to determine whether specific IgE antibodies in serum drawn from sensitized subjects are able to recognize the protein of interest.

- *Assessment of proteolytic stability*. There exists a good, but incomplete, correlation between the resistance of proteins to proteolytic digestion and their allergenic potential, the theory being that relative resistance to digestion will facilitate induction of allergic responses, provided the protein possesses allergenic properties. One approach, therefore, is to characterize the susceptibility of the protein of interest to digestion by pepsin or in a simulated gastric fluid. However, this approach alone may not be sufficient to identify cross-reactive proteins with the potential to elicit allergic responses in food-or latex-sensitized individuals as in the case of oral allergy syndrome or latex-fruit syndrome. Nor are considerations of stability to digestion necessarily relevant for allergens that act through dermal or inhalation exposure and that may have significance for worker health. In these cases, other approaches such as structural homology searches and the use of animal models may be effective in identifying potential new allergens.
- *Use of animal models*. Currently there are no widely accepted or thoroughly evaluated animal models available for the identification of protein allergens. Nevertheless, progress is being made and methods based on the characterization of allergic responses or allergic reactions in rodents and other species have been described.

Although testing strategies for allergens are still evolving and no single test is fully predictive of human responses, the approaches outlined above, when used in combination, allow scientists to address questions of potential allergenicity, and these will increase in precision and certainty with time. Considerations of this type led U.S. federal agencies to deny approval of StarLink corn for human consumption because of the possibility that its Bt protein, Cry9C, may be a human allergen.

This protein had been modified to slow its digestion and prolong its effect in the insect gut and this change rendered the protein less digestible in the human gut as well. After the accidental introduction of StarLink corn into the human food chain, a limited number of illnesses among consumers were reported. These were investigated by the Centres for Disease Control, who found no evidence that the corn products were responsible. However, although this study is reassuring, methodological limitations make it less than conclusive, and it cannot eliminate the possibility that some adverse effects may have occurred that were not reported. Because of this incident, StarLink corn is no longer marketed. With the exception of Cry9C, none of the engineered proteins in foods so far evaluated through the FDA consultation process has had the characteristics of an allergen. The only documented case where a human allergen was introduced into a food component by genetic engineering occurred when attempts were made to improve the nutritional quality of soybeans using a brazil nut protein, the methionine-rich 2S albumin. Allergies to the brazil nut have been documented, and while still in precommercial development, testing of these new soybeans for allergenicity was conducted in university and industrial laboratories. It was found that serum from people allergic to Brazil nuts also reacted to the new soybean. Once this was discovered, further development of the new soybean variety was halted and it was never marketed. This work led to the identification of the major protein associated with Brazil nut allergy, which was previously unknown.

Will Insertion of the Transgene Increase the Potential Hazard from Toxins or Pharmacologically Active Substances Present in the Host?

Concern has been expressed about the randomness with which genes are inserted into the host by current genetic engineering processes. This could, and does, result in pleiotropic and insertional mutagenic effects. The former term refers to the situation where a single gene causes multiple changes in the host phenotype and the latter to the situation where the insertion of the new gene induces changes in the expression of other genes. Such changes due to random insertion might cause the silencing of genes, changes in their level of expression, or, potentially, the turning on of existing genes that were not previously being expressed. Pleiotropic effects could be manifested as unexpected new metabolic reactions arising from the activity of the inserted gene product on existing substrates or as changes in flow rates through normal metabolic pathways.

Unexpected and potentially undesirable pleiotropic or mutagenic changes in the genome of the host do occur, but these would likely be revealed by their effects on the development, growth, or fertility of the host, or by the extensive testing of its chemical composition compared with isogenic untransformed plants, which is a necessary part of any safety evaluation of transgenic crops.

In the U.S., since 1987, the USDA Animal and Plant Health Inspection Service has completed over 5000 field trials with more than 70 different transgenic plant species. The only unexpected result was a mutation in a colour gene and gene silencing through changes in the methylation status of these genes that led to unexpected colour patterns in petunia flowers. Both of these effects are also seen in conventional plant breeding. While the possibility of an undetected increase in a toxic component in a new food cannot be entirely eliminated, the current safeguards make this unlikely, and no toxicologically or nutritionally significant changes of this type are evident in the transgenic plants so far marketed for food production.

Substantial public concern about the safety of BD products was raised in 1989 when a number of cases of eosinophilia-myalgia syndrome (EMS) were reported among users of the amino acid tryptophan as a dietary supplement. By mid-1993, 37 deaths had been attributed to this outbreak. The development of the syndrome appeared among users of some batches of the supplement after a change in the manufacturing process that included the use of a new genetically modified microorganism in the fermentation. However, concomitant with this change were additional alterations in certain filtration and purification steps used previously in the manufacturing process. The exact cause of the outbreak and the nature of the toxic impurity have not been established with certainty.

Thus, it is not possible to determine whether the change in purification, the genetic engineering of the organism, or some other factor or factors were to blame. A subsequent investigation revealed that cases of EMS also occurred among consumers of tryptophan before the GM organism was introduced into the manufacturing process, although at a lower incidence. Thus, the genetic modifications might have caused an increase in the level of the agent that was responsible for tryptophan-associated EMS, but it did not create a novel toxicant. This event is troubling in that the tryptophan would be regarded as highly purified (99.6per cent or higher), and no adequate animal model has been found to replicate EMS, a probable autoimmune disease. This illustrates that toxicology has limits in its ability to explain and predict adverse effects in humans.

These examples indicate that careful analysis of the changes in BD organisms is necessary to ensure against unexpected alterations in the levels of toxins, allergens, and essential nutrients. This analysis will be particularly critical if, as seems likely, engineering of the synthetic pathways of secondary metabolites is undertaken in plants, *e.g.*, to increase their resistance to insects and pathogens or to produce compounds of pharmaceutical value. Such changes might create new and unanticipated secondary compounds with unknown toxic properties. New approaches to profiling changes in metabolites, proteins, and gene expression may be helpful in such cases.

Does the Possible Transfer of Antibiotic Resistance Marker Genes from the Ingested BD Food to Gut Microbes Present a Significant Human Hazard? The development of antibiotic resistance among pathogenic bacteria is a significant human health issue. However, no contribution to antibiotic resistance in gut bacteria arising from antibiotic resistance markers in BD foods has been documented. For several reasons, including the efficient destruction of the resistance gene in the human gut and the very low intrinsic rate of plant–microbe gene transfer, any contribution from this source is expected to be extremely small. Genes for resistance to kanamycin and related antibiotics already occur quite commonly in the environment, including in the flora of the human gut, which naturally contains about 1 trillion (10^{12}) kanamycin-or Neo-mycin-resistant bacteria. Even if the occasional transfer of resistance from plant to bacterium did occur, the practical impact would be negligible. However, since any increase in antibiotic resistance is recognized as undesirable and the technology is now available to omit the use of such marker genes, future genetically modified organisms are unlikely to contain them. Thus, concerns related to their use are likely to diminish.

Will Genetic Transformation Adversely Affect the Nutritional Value of the Host? In the USA, the FDA is entrusted with assuring that the nutritional composition of BD foods is substantially equivalent to that of the Non-modified food. Studies are performed to determine whether nutrients, vitamins, and minerals in the new food occur at the same level as in the conventionally bred food sources. A typical example is the case of Roundup Ready soybeans. In this case, the protein, oil, fibre, ash, carbohydrates, and moisture content and the amino acid and fatty acid composition in seeds and toasted soybean meal were compared with conventional soybeans. Fatty acid compositions and protein or amino acid levels of soybean oil were compared and special attention was given to checking the levels of antinutrients typically found in soybeans, *e.g.*, trypsin inhibitors, lectins, and isoflavones.

One difference between the conventional and Non-conventional soybeans was detected in defatted, Non-toasted soybean meal, the starting material for commercially utilized soybean protein, which is not itself consumed. In this material, trypsin inhibitor levels were 11–26per cent higher in the transgenic soybeans. The levels of the trypsin inhibitors were similar in all lines in the seeds and in defatted, toasted soybean meal, the form used in foods. Except for this difference in trypsin inhibitor levels, all other nutritional aspects were equivalent between the transgenic line and the conventional soybean cultivars. Feeding studies demonstrated that there were no evident differences in nutritional value between the conventional and transgenic soybeans in rats, chickens, catfish, and dairy cattle. Domestic animal feeding studies with a number of other transgenic crops have similarly shown no significant adverse changes in nutritional value.

Will the Transgene Product Adversely Affect Non-target Organisms? In addition to the general concerns addressed that relate to food safety, additional attention is needed when the gene product is pesticidal or otherwise may be toxic to Non-target organisms that consume it. The effects of each transgene product that is designed for pesticidal effects must be evaluated on a case-by-case basis against target and Non-target organisms under specific field growth conditions for each transgenic crop.

The foremost current example of this is the incorporation of Bt genes into crop plants for insect control. The toxic properties of Bt endotoxins to both target and Non-target species of many kinds are well known. They show a narrow range of toxicity limited to specific groups of insects, primarily Lepidoptera, Coleoptera, or Diptera, depending on the Bt strain. Nevertheless,

Bt-producing plants have been tested broadly to determine whether any alteration in this limited spectrum of toxicity has occurred, without the discovery of any unexpected results. Exotoxins and enterotoxins, which are much more broadly toxic than the endotoxins, are also produced by some Bt strains, but these are not present in the transformed plant, because their genes are not transferred into the crop.

In plants transformed with Bt genes to control lepidopterans, toxicity to Non-target lepidopterans would be expected if exposure occurs by feeding on the transformed crop. Particular concern has been expressed over the potential toxicity of the Bt toxin in corn pollen to the Monarch butterfly after initial laboratory studies showed increased mortality in larvae fed on leaves dusted with transgenic pollen. However, most transgenic corn pollen contains much lower Non-lethal levels of Bt toxins than the strain used in this study, and there is only a limited synchrony between the feeding period of the most sensitive younger larvae and the period when corn pollen is shed.

Also, corn pollen does not typically move far beyond the borders of the field, leaving significant amounts of milkweed uncontaminated in many locations. For these reasons, a detailed risk assessment concluded it is unlikely that a substantial risk to these butterflies exists in the field since only a negligible portion of the population is exposed to toxic levels of Bt. Beyond the question of the potential toxicity of Bt corn to such valued insects, it is also important to recollect that the common alternative is to spray corn with synthetic insecticides, which are not as selective as the Bt toxin. In a sweet corn field containing milkweed plants and treated with a synthetic pyrethroid for insect control, 91–100per cent of the monarch butterfly larvae placed on the milkweed leaves after spraying were killed. In plots where Bt sweet corn was planted and the pollen fell naturally on the milkweed leaves, larval death rates were much lower (7–20per cent) and indistinguishable from those in untreated non-Bt corn plots.

GENETIC MATERIAL IN PRODUCT CHARACTERIZATION

Product characterization takes into consideration information relating to the modified food crop, the introduced genetic material and its expression product, and acceptable levels of inherent plant toxicants and nutrients. All characteristics of the gene insert must be known, including the source(s), size, number of insertion sites, promoter regions, and marker sequences. It must be established that the transferred genetic material does not come from a pathogenic source, a known source of allergens, or a known toxicant-producing source. The introduced genetic material should be well characterized to ensure that the introduced gene sequences do not encode harmful substances and are stably inserted within the plant genome to minimize any potential opportunity for undesired genetic rearrangement.

Analytical data are required to evaluate the nutritional composition, the levels of any known toxicants, anti-nutritional and allergenic substances, and the safety-in-use of antibiotic resistance marker genes. Any new substances introduced into crops through rDNA technology (*e.g.* proteins, fatty acids, carbohydrates) will be subject to pre-market review as food additives by the US FDA, unless substantially similar to substances already safely consumed as a part of foods, or that are considered GRAS. To date, substances that have been added to foods through rDNA technology have been previously consumed or have been determined to be substantially similar to substances already consumed as a part of the diet. As such, introduced substances have been considered exempt from the requirement for pre-market approval as food additives with the US FDA.

A more rigorous safety evaluation of a GM crop is warranted if the introduced gene sequence(s) has not been fully characterized, the nutritional composition has been significantly altered, antibiotic resistance marker genes have been used during its development, or if an allergenic protein or toxicant has been detected at levels higher than what is typically observed in edible varieties of the same crop species. In any event, determinations as to the safety of substances that have been introduced into new plant varieties through rDNA technology are made on a case-by-case basis.

Although an evaluation of the introduced gene sequences and expression product(s) provides assurance as to their safety, further studies may be required to predict whether unexpected effects may result following their interaction with other genes within the plant. In addition, the product characterization of a GM plant involves assessing sequence homology to known toxicants and allergens, thermal and digestive stability, and if required, the results of both *in vitro* and *in vivo* assays to demonstrate lack of toxicity.

COMPOSITIONAL ANALYSIS

The results of field trials performed over several years serve to characterize the phenotypic and agronomic characteristics exhibited by the plant (*e.g.* height, colour, leaf orientation, susceptibility to disease, root strength, vigour, fruit or grain size, yield, etc.), as well as to provide the materials required for the compositional analysis. Any anomalies in the phenotypic or agronomic characteristics exhibited by a plant may result in a requirement for additional information. Protein, fat, fibre, starch, amino acid, fatty acid, ash and sugar levels are determined, as well as the levels of anti-nutrients, natural toxicants or known allergens. Studies of the nutritional composition are performed to determine whether the levels of any key nutrients, vitamins or minerals have been altered as a result of the genetic modification.

Based on the results of these studies, a determination is made as to whether the phenotypic and agronomic characteristics of a GM crop or the concentrations of inherent constituents fall within ranges typical of its conventional counterpart. If inserting a new gene causes no change in any of the assessed parameters, the US FDA can conclude with reasonable assurance that the GM crop is as safe as the conventional crop. If the levels of essential nutrients or inherent toxicants are found to be significantly different in the GM crop, the US FDA may recommend additional action prior to commercialization, such as obtaining food additive status, or the use of specific labels to alert consumers of an altered nutritional content, etc.

ALLERGENICITY

In consultation with scientific experts in the areas of food safety, food allergy, immunology, biotechnology and diagnostics, the US FDA published guidelines for assessing the allergenicity of GM foods or food components in 1994.

The approach to assessment is multi-faceted, incorporating data regarding the origin of the genetic material, and the biochemical, immunological and physicochemical properties of the expressed protein. The overall assessment is reliant upon the fact that all known food allergens are proteins and, notwithstanding the number of shared properties between allergenic and non-allergenic food proteins, food allergens tend to exhibit a number of similar characteristics. In general, food allergens share a number of common properties: they have a molecular weight of over 10 000 Da; they represent more than 1per cent of the total protein content of the food; they demonstrate resistance to heat, acid treatment, proteolysis and digestion; and they are recognized by IgE.

For gene sequences derived from known allergenic sources (*e.g.* peanuts), the developers of GM plants are expected to demonstrate that allergenic proteins have not been introduced into the food. For assessment purposes, it is assumed that any genetic material derived from a known allergenic source will encode for an allergen. To demonstrate otherwise, the amino acid sequence of an expressed protein must be compared with that of known allergens using protein sequence databases. Furthermore, *in vitro* and/or *in vivo* immunologic analyses using the sera of allergic patients sensitive to the source of the genetic material may need to be performed to determine whether or not a potentially allergenic protein is being expressed in the GM food.

Some GM foods may be modified to express genes from a source that is not known to be allergenic when consumed. Under these circumstances, the US FDA follows a similar decision tree-based approach to determining the allergenic potential of the expression product. In assessing these proteins, should any amino acid sequence exhibit homology with a known allergen, the expressed protein would then be evaluated in immunologic tests using the sera of patients known to be allergic to the identified homologous protein. Regardless of the origin of the genetic material, physicochemical studies are performed *in vitro* to provide information concerning the expected stability of the expressed protein. All known food allergens tend to be resistant to digestive degradation, as demonstrated in simulated gastric fluid models, or to decomposition under conditions of food processing.

In making a determination regarding a GM food, it is the totality of the biochemical, immunological and physicochemical properties of the introduced protein that provides guidance as to the allergenic potential of such a protein being expressed in food. The level of protein expressed, produced and consumed as a part of the diet is also a primary indicator of the allergenic potential, since nearly all food allergens are known to be major proteins in their respective foods. If the results of any of these studies suggest an allergenic potential, the US FDA may recommend further scientific evaluation, require special labelling to alert sensitive consumers, or alternatively caution the developer about proceeding with the development of a particular GM food.

Most recently, a joint FAO/WHO Expert Consultation on Allergenicity of Foods Derived from Biotechnology recommended a revised decision tree. This decision-tree strategy was modified from the previous version FAO/WHO to include a revised definition of sequence homology for gene product comparison; a greater emphasis on serum testing, even with gene products without homology to known allergens and not derived from an allergenic source; and animal models to assess potential allergenicity, despite acknowledgement by the Expert Consultation that these models are currently under development and at present, not predictive of food allergies in humans.

More recently, the *ad-hoc* Open-Ended Working Group on Allergenicity established by the *ad-hoc* Intergovernmental Codex Task Force on Foods Derived from Biotechnology, considered the FAO/WHO strategy in drafting an approach to assessing the potential allergenicity of foods derived from rDNA plants.

The Codex Working Group recognized the absence of a definitive predicitve test for allergenicity in humans to a newly expressed protein and recommended an integrated, stepwise assessment strategy. The strategy recommended by the Working Group, and accepted by the Codex Task Force for inclusion in the draft forwarded for final adoption by the Codex Alimenarius Commission, is consistent with that of FAO/WHO. However, the Working Group suggested that some of the modifications included in FAO/WHO could contribute to the overall weight of evidence of any conclusion of potential allergenicity (*e.g.*, allergen specific serum depositories, animal models), pending development and validation.

ANTIBIOTIC RESISTANCE

The use of antibiotic-resistance genes as selectable markers has been common practice in the development of new plant varieties using rDNA technology. Concerns relate to the potential transfer of antibiotic-resistance genes from GM plants to pathogens in the environment or to the gut of humans consuming foods or food components. Issues relating to the use of antibiotic resistance genes were identified in the 1992 US FDA Statement of Policy Statement as well as discussed in additional guidance entitled *Guidance for Industry: Use of Antibiotic Resistance Marker Genes in Transgenic Plants*. The guidance provided within these documents was established in consultation with experts in the fields of microbiology, medicine, food safety, bacterial and mycotic diseases, and includes suggestions with respect to the continued safe use of antibiotic-resistance marker genes by the developers of new plant varieties.

The use of marker genes that encode resistance to clinically important antibiotics has raised questions as to whether their presence in food could reduce the effectiveness of oral doses of the antibiotic or whether the gene present in the DNA could be transferred to pathogenic microbes, rendering them resistant to treatment with the antibiotic. The risk of transfer of antibiotic-resistance genes from plants to microorganisms considered to be pathogenic to humans, however, is considered to be minimal if not insignificant. Furthermore, the potential risks are becoming less of a concern as more developers are beginning to research the use of alternative technologies (*e.g.* non-resistance-based markers) in plant breeding. The conclusions with respect to the safe use of antibiotic-resistance marker genes are consistent with the findings of other national and international food safety organizations.

Consultation and Filing Process

The submission of a Pre-market Biotechnology Notification (PBN) has recently been proposed as a mandatory requirement for the commercialization of bioengineered foods and food ingredients within the United States. Minimal differences exist between the information to be submitted as a part of a PBN and that previously presented in voluntary consultations with the US FDA. As proposed, developers are still being encouraged to consult with the US FDA as early and as often as necessary in the development of a bioengineered food, such that any potential scientific or regulatory concerns can be identified and addressed prior to the submission of the PBN.

Guidelines for performing consultations with the US FDA were released in a publication entitled *Guidance On Consultation Procedures: Foods Derived from New Plant Varieties* in 1997. The publication recommended an approach for developers to proceed by submitting a request for consultation, outlined the internal process by which all requests would be handled, and included additional guidance as to the type of safety and nutritional information to be presented during consultation with the US FDA. Once sufficient safety and nutritional information has accumulated to demonstrate that a product is safe and in compliance with the FFDCA, developers typically schedule a consultation to present their scientific findings and conclusions to the US FDA. Consultations prior to notification not only serve to keep the US FDA informed of advances made in the application of rDNA technology in food production, but also keep the developers of bioengineered foods aware of emerging safety, nutritional or regulatory concerns of the US FDA.

Consultations are considered complete when all safety and regulatory concerns between the US FDA and the developer have been resolved. The US FDA proposes to perform an initial evaluation of a PBN within 15 days of receipt to determine completeness, at which point, if considered complete, the PBN will be filed, and a response can be expected within 120 days. The US FDA does not issue a product approval *per se*, but informs the developer by letter that:

- The evaluation period has been extended;
- The notice is not complete and why;
- It has no further questions 'at this time' based on the information that has been presented.

Labelling

The FFDCA defines what information must be disclosed to consumers on a food label, such as the common or usual name, and other limitations concerning the representations or claims that can be made or suggested about a food product. All foods must be labelled truthfully and not be misleading to consumers. Taking this into consideration, the FFDCA does not stipulate the disclosure of information on the basis of consumer desire to know. Labelling may be considered misleading if it fails to reveal material facts in light of representations that are made with respect to a product.

The labelling of foods derived from new plant varieties, including plants developed using rDNA technology, was originally addressed in the 1992 US FDA Statement of Policy, and most recently discussed in Draft Guidance for Industry for the voluntary labelling of bioengineered foods.

To date, the US FDA is not aware of any information that would distinguish foods developed using rDNA technology (*e.g.* bioengineered foods) as a class from foods developed through other methods of conventional plant breeding, and as such, have not considered the method of development a material fact requiring disclosure on product labels. Nevertheless, after extensive consultation, including thousands of written comments and a series of public meetings, the US FDA has observed 'a general agreement that providing more information to consumers about bioengineered foods would be useful'.

Requirements for Labelling

Special labelling is required if the composition of the bioengineered food differs significantly from its conventional counterpart. For example, for a food that has been genetically modified to contain a new major sweetener, a new common or usual name or other labelling may be required. Similarly, if a GM food contains an allergen that consumers would not expect to be present in that food, special labelling may be necessary to alert sensitive consumers. If a protein commonly associated with an allergic reaction (*e.g.* peanut protein) is transferred to another food through genetic modification, the US FDA would evaluate whether labelling would provide sufficient consumer protection. If labelling would not be considered to provide a sufficient level of protection, the US FDA would take appropriate steps to ensure the GM food would not be marketed. Therefore, current policy requires a GM food to be labelled when the resulting product poses a safety issue or is substantially different from its conventional counterpart, and as a result, could be considered to pose a misrepresentation to consumers.

The 1992 US FDA Statement of Policy and the recent Draft Guidelines do not consider the use of rDNA technology in the development of food products to be a material fact requiring specific disclosure on the label. Rather it is a method of development, similar to other methods of plant breeding, which have not required disclosure on the label. Bioengineered foods cannot be distinguished compositionally from foods modified through more conventional methods and thus do not require specific disclosure through labelling. The Draft Guidelines reaffirm the US FDA position that bioengineered foods do not require special labelling.

Voluntary Labelling

To provide guiding principles for voluntary labelling, in recognition of the desire of certain manufacturers to label foods as produced either with or without bioengineering, the US FDA published a 'Draft Guidance for Industry: voluntary labelling indicating whether the foods have or have not been developed using bioengineering'. Emphasizing that the use of rDNA technology 'is not a material fact', the US FDA recognizes that some consumers want disclosure of bioengineered content and that some manufacturers wish to provide it. In response, Draft Guidance was issued with suggestions concerning the use of labelling statements that are not considered misleading.

PRESENCE OR USE OF RDNA

The US FDA provided several examples of how disclosure of bioengineering can be accomplished, be informative and not be misleading.

- *Example* 1 'Genetically engineered' or 'This product contains corn meal that was produced using biotechnology'. These disclosures reveal the minimum amount of optional information about bioengineering.
- *Example* 2 'This product contains high oleic acid soybean oil from

soybeans developed using biotechnology to decrease the amount of saturated fat.' This statement explains the proper and required name of this type of soybean oil, since it differs from what would be considered as soybean oil. The optional comments about biotechnology and decreasing saturated fat provide information that could be seen as benefits of the product.

- *Example* 3 'These tomatoes were genetically engineered to improve texture.' This example was one included to illustrate how it could be misleading to consumers if they cannot discern a difference in texture, but would not be misleading if they can tell a difference. If the former, and the new texture is to facilitate processing, then this intention should be made clear, *i.e.* 'to improve texture for processing'.
- *Example* 4 'Some of our growers plant tomato seeds that were developed through biotechnology to increase crop yield.' This is another example of optional information that would explain an indirect, agricultural benefit.

ABSENCE OR NOT BIOENGINEERED

The US FDA provides an important commentary in the Draft Guidance about 'genetically modified organisms (GMO)' versus 'bioengineered'. It would be technically inaccurate to use the phrase 'not genetically modified' or 'GMO-free' to mean that bioengineering was not used. This is because most conventional foods have been genetically modified over the years by traditional crop breeding practices. Examples of acceptable voluntary statements would include:

- We do not use ingredients that were produced by biotechnology.'
- This oil is made from soybeans that were not genetically engineered.'
- Our tomato growers do not plant seeds developed using biotechnology.'

Another important point is made about a term such as 'GMO-free' that is misleading for two reasons:

- Many foods do not contain organisms anyway and therefore should not be labelled as 'organism-free';
- Free' implies complete absence or 'zero' amount of bioengineered material.

In a practical sense, it is impossible to demonstrate analytically the complete absence of anything. Therefore, in reality, a threshold for possible adventitious presence of a low level of bioengineered material may be necessary and has been the subject of much debate. The Agency also provides additional guidance about misleading statements that:

- Could be interpreted to suggest that the absence of bioengineering would make the food superior to the bioengineered alternative;

- Claim the absence of one bioengineered ingredient when the food contains another ingredient that is bioengineered;
- Claim that a food is not bioengineered when in fact this type of food (*e.g.* green beans) has never been modified through rDNA technology.

CONCLUSION

Genetically modified organisms (GMOs) are a fact of modern agriculture, and are here to stay. GMOs are also a fact of public preoccupation and opinion, which politicians must take into account. FAO recognizes the great potential and the complications of these new technologies. We need to move carefully, with a full understanding of all the factors involved. In particular, we need to assess GMOs is terms of their impact on food security, poverty, biosafety, and the sustainability of agriculture. Will GMOs increase the amount of food in the world, and make more food accessible to the hungry?

Clearly, GMOs should be seen not in isolation as technical achievements. Hence, we will discuss not the specifics of GMO technology, but the context in which they are developed and deployed, and about how public opinion and government policy on GMOs are formed.

The public in many countries distrusts GMOs. They are often seen in the context of globalization and of privatization and even as "antidemocratic" or "meddling with evolution". There are as yet few perceived advantages for the public, because GMO applications to date have concentrated on reducing costs for producers without direct consumer benefits.

In particular, it has been a tactical error of the industry to concentrate on pesticide-resistance as one of the earliest applications, as this has stimulated environmental concerns. The public often confuses the industry with the science. And consumers worry about risk, not about scientific freedom. Scientists in both the private and public sectors clearly see genetic modification as a major new set of tools. They are also participants and spectators in a major shift of research from the public to the private sector, which will undoubtedly influence the future direction of research and research investment. As shareholders in the GMO debate, scientists must recognize that there is also a substantial public distrust of science.

SOCIETY OF TOXICOLOGY

The Society of Toxicology (SOT) is committed to protecting and enhancing human, animal, and environmental health through the sound application of the fundamental principles of the science of toxicology. It is with this goal in mind that the SOT defines here its current consensus position on the safety of foods produced through biotechnology. In this context, biotechnology is taken to mean those processes whereby genes that are not endogenous to the organism (transgenes) are transferred to microorganisms, plants, or animals employed in food production, or where the expression of existing genes is permanently modified, using the techniques of genetic engineering. We intentionally avoid using the term genetically modified organisms (GMOs) or foods in this context, since conventional techniques of plant and animal breeding, which are not considered here, also involve genetic modification. The extent of the genetic changes resulting from such conventional breeding techniques, which is generally undefined, far exceeds that typically produced by transgenic methods. Consequently, it is important to recognize that it is the product, and not the process of modification, that is the focus of concern regarding the human or environmental safety of biotechnology-derived (BD) foods.

The principal responsibilities of toxicologists are to define and characterize the potential for natural and manufactured materials to cause adverse health effects and to assess, as accurately as possible, the plausibility and level of risk for human or animal health or for environmental damage under a defined set of circumstances. It is not the task of the Society of Toxicology to determine the overall value of a product or process by balancing health or environmental risks with potential benefits, or to choose between different strategies to manage risk, although toxicological considerations are important in both processes. Our purpose here is rather to identify and consider the primary toxicological issues associated with BD foods. Major areas of concern in the development and application of such foods in agriculture relate to the possibility of deleterious effects on both human health and the environment. We do not consider here some aspects of the possible environmental impact of GM organisms such as gene transfer to nonengineered plants.

GENETICALLY ENGINEERING IN PLANTS AND FOODS

The genomes of all eukaryotic species consist of single-copy, middle repetitive and high copy number sequences. To gain an understanding of the per centages of each of the classes within any particular species, a group of random clones can be hybridized to blots of plant DNA. One such study was performed by Zamir and Tanksley. They hybridized 50 random genomic clones to tomato DNA. Washing was performed at two stringencies. The following are the results with regards to copy number.

Clone Class	Low Stringency Wash	High Stringency Wash
Single copy clones	44per cent	78per cent
Multiple copy clones	46per cent	18per cent
Repetitive clones	10per cent	4per cent

The table shows that hybridization stringency has a significant effect upon the number of sequences to which a sequence hybridizes. At the higher stringencies, most of the clone recognized only a a single copy within the genome. What this also shows is that the tomato genome contains many sequences that are about 80per cent homologous, but fewer sequences that are highly homologous. Clearly similar sequences must have diverged by some mechanism during the evolution of the species. The authors also hybridized these 50 clones to filters containing tomato, related tomato species, eight Solanum species and one member of the Curcurbitaceae family.

As a control,single copy tomato cDNA clones were also hybridized to these clones. At moderate stringency, the cDNAs hybridized to all the tomato species, and most hybridized to the Solanum species. 80per cent of the clones hybridized to the related Curcurbitaceae species. In contrast, only 50per cent of the random clones hybridized to related tomato and Solanum species, and only 10 per cent hybridized to the Curcurbitaceae species. The principal conclusion that can be drawn from these hybridizations is that the random sequences (as represented by the random genomic clones) are evolving faster than the single-copy sequences. Why? At the higher stringency, homology to distant related species was reduced. Therefore these sequences have undergone greater divergence. This evidenced by the fact that 0/5 of the repetitive sequences hybridized to tobacco, whereas 20/45 of the single and multiple copy clones hybridized to tobacco.

COLLECTION OF CULTIVARS OR SPECIES

The typical experiment begins with a large collection of cultivars or species. These samples are then analysed by hybridization with RFLP clones or more recently by the PCR-RAPD amplification. The similarity of each pair of samples is measured by calculating the number of common bands or amplification products. One estimator was developed by Nei and L. The formula is:

$$F = 2nXY/(nX + nY)$$

where nX and nY are the total number of fragments for sample X and Y and nXY is the number of fragments shared by the samples. Data is normally collected for a relatively large number of hybridizations or PCR-RAPD amplifications. The similarity data is then used in a cluster analysis to develop dendrograms which show the molecular relatedness of the species.

These types of analyses can:

- Provide independent support to previous phylogenetic and evolution hypotheses.
- Identify gene pools within a genus.
- Estimate genetic diversity within a genus.
- Provide necessary data to select appropriate parents for a molecular mapping project.

PHYSICAL AND GENETIC DISTANCES

All distances found on a linkage maps are the product of genetic recombination and therefore are considered to be genetic distances. Two pairs of loci that are genetically the same distance apart may not be physically the same distance apart because of suppressed recombina-tion in the region between two of the loci. Those loci where suppression of recombination does occur will physically be farther apart. RFLP hybridizations to large fragments of DNA are necessary to accurately gauge the physical distance.These experiments are similar to other Southern hybridizations.

The DNA is cut with restriction enzymes first. The choice of enzymes is important because you want to generate large fragments. For these experiments you want to cut the DNA with enzymes that recognize 8 nucleotide sequences in the target DNA. Because these sites will be rare (every 65,536 bases on average) the digestion products will range from several hundred kilobase to several megabases. Because the normal gel systems do not separate large fragments very easily, it is necessary to run pulsed-field-gel-electrophoresis. This technique was first applied to the separation of yeast chromosomes and was quickly applied to large-scale mapping experiments.

Physical and genetic distances are then resolved by hybridizing clones which define closely linked genetic markers, to DNA that has been cut with rare cutting enzymes. What you are searching for is co-hybridization of the two clones to the same restriction fragment.

If two clones hybridize to the same fragment, then the maximum distance between those two clones is the size of the restriction fragment. Because you already know the genetic distance between the loci, you can correlate the genetic and physical distances in this region. The following table give the relationship between the physical and genetic distance in three species.

Species	kilobases/centimorgan
Arabidopsis	139
Tomato	510
Corn	2140

But remember, these values are estimates of a single region of the genome of each species, and it is quite common to see that two regions of the same species have different amount of DNA per genetic distance. Correlations between physical and genetic distances have also been performed using deletion stocks in wheat.

A series of deletions stocks of a specific wheat chromosome were analysed with a probes known to hybridize to that chromosome. "Each locus was assigned to the chormosome region between the breakpoint of the largest deletion where the band was present and the next larger deletion where the band was absent." (Werner *et al*. PNAS 89:11307–11311) The authors noted large discprencies between the physical and genetic maps of chromosomes 7B and 7D.

Two loci that are located near the centromere of 7B are 7 cM apart genetically, yet the distance between the two loci spans 25per cent of the chromosome. Another region at the distal region of long arm of 7B which accounts for about 15 per cent of the chromosome is 91 cM long genetically. This points out the difference of recombination that can occur within a single chromosome.

COMPARATIVE GENOME MAPPING

Often the clones used to identify RFLPs in one species can be used in a second species. These heterologous probes can then be ued to develop a RFLP map in the second species. Once the two species have been mapped, the relative evolution of the two species can be compared. Comparative mapping can identify inversions, translocation and duplications that have occurred. Genetic factors can also be assessed by comparing map distances of genes with conserved gene order in the two species.

A comparison of the related grass species sorghum and maize was made using maize clones. The diploid chromosome number of the two species is twelve. Many of the sorghum chromosomes contained regions from two of the maize chromosomes. This result may represent the ancestral duplication of chromosomal material. Furthermore, during maize evolution duplicate genes have also occurred to a greater extent than that seen in sorghum. Only nine inversions of gene order have been observed between the species. Maize and sorghum were next compared with regard to genetic distance. Conserved gene orders were compared, and these linkages measured 862 cM in maize and 835 cM in sorghum. Therefore, it can be concluded that since the divergence of sorghum and maize, large chromosomal changes have not occurred and much of the recombination distance has been maintained.

The same type of analysis was performed with two more distant species, tomato and pepper. Little linkage conservation appears to have been maintained since the divergence of these two species, even though the chromosome number has been conserved. The largest conserved linkage block is a 63 cM block of tomato chromosome two that was located on chromosome I of pepper. Some chromosomes of pepper contained six distinct regions of the tomato genome (pepper chromosome X). These chromosomal breakages and rearrangements are not centromeric in nature, suggesting that evolution involved breakage throughout the genome. Because duplicated regions were found in pepper, it was concluded that the gene duplications events within pepper occurred after divergence from tomato. Linkage distances of conserved gene orders were quite similar.

GOAL OF PLANT BREEDING

One goal of plant breeding is to introduce a gene from a donor parent to improve a cultivar for a specific trait. For example, a wild germplasm could be used as a source of disease resistance. At the same time the breeder does not want to carry any of the other genes from the wild germplasm that might reduce the a agronomic fitness of the cultivar. The backcross method of plant breeding is one manner in which the introduction of a specific gene is accomplished. One genetic feature though of backcross breeding is linkage drag. This refers to the reduction in fitness in a cultivar due to deleterious genes introduced along with the beneficial gene during backcrossing. Molecular makers offers a tool in which the amount of wild or alien DNA can be monitored during each backcross generation.

First, though, lets look at the amount of alien DNA that can be maintained after a backcrossing programme. The accompanying figure handed out in class shows the chromosomal region around the Tm-22 allele in a number of tomato cultivars. This allele was introduced from the wild tomato species L. peruvianum, and it provides resistance to tobacco mosaic virus. Large variation in the amount of introgressed DNA was observed. For example, Craigella-Tm-22 contains 51 cM of wild DNA, whereas Vendor-Tm-22 and Nova-Tm- 22 each contain about 8 cM of introgressed DNA.

Backcrossed lines could be developed rapidly in which only a small amount of foreign DNA is linked to the gene of interest. After the first backcross all lines with the gene of interest could be screened with an RFLP that is 1 cM away. Those lines in which a crossover occurred at this marker would be selected. Then those lines would be crossed to the recurrent parent.

The progeny from these crosses would then be scored for a second marker 1 cM away on the other side of the gene. Again crossovers progeny could be selected. Thus, in only two generations the amount of wild DNA could be reduced to 2 cM. The key to this procedure is to have markers that are polymorphic between the two parents that are closely linked to the gene of interest. Young and Tanklsley applied these principles to the reduction of L. peruvianum DNA in Craigella-Tm-22. The authors were able to reduce the amount of DNA on one side of the Tm-22 allele from 47 to 7 cM in one generation by selecting for crossovers at the CD32A locus.

MAPPING MARKERS

Traditional quantitative genetic research defined a quantitative trait in terms of variances. The total phenotypic was first partitioned into genetic and environmental variances. The genetic variance could then be further divided into additive, dominance and epistatic effects. From this information it was then possible to estimate the heritability of the trait and predict the response of the trait to selection. It was also possible to estimate the minimum number of genes which controlled the trait.

Mapping markers linked to QTLs identifies regions of the genome that may contain genes involved in the expression of the quantitative trait. But what functions could these genes be encoding. To answer this question we should consider a trait such as yield. What types of qualitative genes (genes inherited as simple genetic factors) could be involved in the expression of yield? The first event required for yield is meiosis. Therefore any gene that is involved in gamete formation could potentially be considered a QTL. Any of the genes involved in the protein and carbohydrate biosynthetic pathways could also affect the final yield of a plant and could also be considered to be QTLs. The markers associated with a QTL each account for only a portion of the genetic variance. Likewise each of these genes of known function may only account for a portion of the final yield. An important question that can now be posed is whether any known genes map as QTLs. Beavis *et al.* analysed four populations of maize and found molecular markers linked to plant height. No marker was consistently associated as a QTL with plant height in all four populations. Each of the ten maize chromosomes contained a marker linked to a QTL for at least one of the four populations. The authors further were able to demonstrate that a number of the QTLs identified by the molecular markers mapped to regions containing genes known to have a qualitative effect on plant height.

For example, on chromosome 9 the gene d3 resides within 10 cM of a plant height QTL. This gene is involved in gibberellic acid sensitivity. Mutants do not respond to the hormone and do not undergo the normal cell elongation. These mutants are phenotypically shorter than normal maize plants. The question that needs to be raised is if the QTL that was being identified by the molecular marker is actually the d3 gene. It could be possible that what is actually being measured by the marker is the linkage of the marker with the gene. The statistical analysis of quantitative traits provided valuable information for the plant breeder. Molecular analysis of quantitative traits now provides new tools, not only as selection tools for plant breeding, but as starting points for the cloning of these genes. These objectives could not have been realized without molecular markers.

PLANT GENOME ORGANIZATION AND STRUCTURE

Eukaryotic genomes are much more complex than prokaryotic genomes. And further, plant genomes are more complex than other eukaryotic genomes. Prior to the development of recombinant DNA technology genomes, were analysed by reassociation kinetics techniques. Reassociation kinetic experiments are performed by melting DNA and allowing it to reanneal upon itself or with another population of either DNA or RNA molecules. The kinetics of the reassociation provide data that can be used to analyse the overall structure, evolution and expression of genomes.

The most common method of denaturing duplex DNA is by heating to 100°C. But how can we monitor this denaturation and subsequent renaturation? The most common method is by measuring the absorbance change in the ultraviolet region at 260 nm. The important property is that melted, single-stranded DNA absorbs about 40per cent more at 260 nm than duplex DNA.

If we slowly heat DNA the absorbance will increase dramatically over a short range of temperatures. The mid-point of this transition is called the melting temperature or Tm. Under physiological conditions, the Tm usually lies in the range of 85-95°C. Thus, without altering the cellular conditions, the duplex DNA is stable in the cell. The exact temperature that a particular DNA melts depends on several parameters. The GC content is important because GC base pairs have three hydrogen bonds compared to the two for AT base pairs. Thus, the higher the GC content the higher the Tm.

Table. Effect of GC Content on Tm

per centGC	Melting Temperature
40	87°C
60	95°C

DNA denaturation is reversible, and the reversible process is called renaturation. This process requires that the temperature be lowered gradually.

As this lowering of temperature proceeds the following occurs:

- Single-stranded molecules randomly encounter each other.
- Short complementary stretches of duplex DNA form.
- The DNA then zips back together to form the original structure.

Obviously if a single molecule is denatured it should be able to reform completely. But if two separate molecules are denatured together, for example wheat and barley DNA, then complementary regions between the two DNAs will be able to form duplex DNA. The ability of two molecules to renature is called hybridization.

Hybridization: The pairing of complementary nucleic acids. Performing hybridization experiments in a solution is called liquid hybridization. We have already discussed the related procedure called filter hybridization.

DNA MOLECULE

If a DNA molecule is melted and allowed to reassociate, the complexity of the genome dictates the rate in which duplex DNA will form. If we consider a simple molecule that consists of alternating GCs, this molecule will be able to form a duplex quicker than a molecule that consists of repeating blocks of AGCT. As the number of different combinations of bases increases, the time required for complete duplex formation to occur will increase.

Renaturation, or duplex formation requires random collisions between two single-stranded molecules. This process follows second-order kinetics and is concentration dependent. We will not go through the derivation of the formula but the important parameter used to define a certain DNA is:Cot½. This value is defined as the amount of time required for one-half of the DNA to reanneal or form duplex DNA. The units for this parameter is moles of nucleotides per liter per second. The more complex the genome of interest, the longer it will take for like sequences to reanneal. Consequently, the Cot½ will be larger. Thus in terms of reassociation kinetics complexity has a specific definition.

Complexity: The total length of different sequences. For example, E. coli is considered to have a complexity of 4.2 × 106 base pairs. What is the experimental procedure used to derive these values?

In general the procedure is:

- Shear the DNA to be analysed to a length of about 300 bp.
- Melt the DNA (usually in 0.12 M phosphate buffer) by boiling for 5 min.
- Quickly place at 60°C.
- Take aliquots at different time points. Separate single-stranded DNA from double-stranded DNA by hydroxyapatite. Measure the amount of DNA that is double-stranded by absorbance at 260 nm.
- Plot the amount that is single-stranded versus the Cot value. The Cot value is expressed in log equivalent. This plot depicts the Cot curve.

When this type of experiment is performed with eukaryotic DNA three components are usually seen. These components each reanneal with their own unique Cot½ value. The three components are termed the fast, intermediate, and slow components. Why do we see these three components? Eukaryotic genomes are characterized by sequences that are represented by different copy numbers. If a sequence is found many times in the genome, it will reanneal much quicker than those sequences that are found only once in the same genome. Thus the equivalent Cot curve for a eukaryotic genome will be different than a genome, such as E. coli, which only contains single copy sequences.

A comparison of the Cot value of each of these components with an E. coli standard allows us to derive the complexity of each component. The complexity of the slow component of the genome is greater than that for the other two components and is considered to represent the single copy portion of the genome. The complexity of the slow component can be used as a good estimate of the genome size.

The genome size will be the sum of the lengths of all the unique sequences. Using the example from Genes V - Lewin, p.664, the complexity of the slow component is 3×10^8 bp and the complexity of the intermediate component is 6×10^5. If we divide the complexity of the slow component into the intermediate component we get 2×10^{-3}. This demonstrates that the intermediate component contributes very little to the complexity of the genome. Therefore, the complexity of the slow or single copy portion of the genome can be considered equal to the genome size.

To derive the complexity of each component, it is necessary to run a standard, such as E. coli DNA, with each experiment. *E. coli* is considered to consist of only single- copy sequences. Let's say that in the experiment from which the Cot for each component was derived, the Cot½ value for E. coli was

4. Experimentally it was determined that the slow component comprised 45per cent of the total DNA. Therefore, if only that component was annealed, the Cot½ value would be 283 (630 × 0.45). That value is 71 (283/4) times slower than for E. coli. Therefore the complexity of the slow component is 71 times that of E. coli or 3.0×10^8 ($71 \times 4.2 \times 10^6$). The complexity of the other components is derived similarly. Genome size is quite variable throughout the biological world and the genome size in plants shows the greatest variation of any kingdom in the biological world.

Table. Variation in Genome Size among Plants

Species	kb/haploid	pg/haploid
Arabidopsis	7×10^4	0.15
Lily	1×10^8	100.00

Conversion factor: 1 pg = 0.965 × 109 bp = 6.1×10^{11} daltons

One manner in which a genome can be described is by determining the distribution of fast, intermediate and slow components in the genome. For comparison purposes, what does the human genome look like? Distribution Sizes Among Components of the Human Genome

Component	per cent of Genome
Fast	6
Intermediate	38
Slow	50

The table Sequence Distribution of Selected Plant Species lists the different components from different plant species. The plant species exhibit a wide range of values for each of the components. The genome of Arabidopsis is essentially entirely single copy sequences (the repetitive sequences have been determined to be essentially all chloroplast DNA). At the other extreme, pea and wheat genomes have only 10–20per cent single copy sequences. Reassociation kinetic experiments of polyploid species, such as bread wheat (Triticum aestivium) were unable to derive a component that displayed true single copy kinetics. Instead the slowest component appeared to act as if it consisted of copies represented three times. This result is consistent with the current hypothesis that bread wheat was developed from the introgression of three diploid wheat species. Genomic analysis suggest that this hypothesis is correct since the slowest component appears to consist of sequences represented three times.

ORGANIZATION OF SINGLE-COPY SEQUENCES

Single-copy sequences are interspersed throughout the plant genome. These sequences are bounded by repeat sequences. The length of the single-copy regions varies widely among plant species.

In general, two types of arrangements are recognized:

- *Short Period Interspersion*: Single copy sequences of 300-1200 bp are interspersed as islands among short lengths of repeat sequences

- *Long Period Interspersion*: Single copy sequences of 2000-6000 bp are interspersed as islands among repeat sequences

How can the interspersion type be determined? The reassociation kinetic experiments are performed with 300 nt long fragments. These experiments give a characteristic Cot curve that defines each of the components. But what happens if the fragment that is followed is of a longer length, for example 900 nt? This fragment could be considered to consist of three 300 nt fragments, and each fragment may be from any of the three components.

Indeed this is how these experiments may be performed. A tracer of a longer length (such as 900 nt) is prepared and radiolabelled, for example with H ^ 3. This tracer is added to a normal Cot reaction in such a low concentration that the rate is not affected. The reaction is then allowed to proceed to a Cot value determined from a normal experiment with only 300 nt fragments where the repetitive sequences have reannealed, but the single-copy sequences have not.

Then the amount of tracer which remains single-stranded is determined. What will be seen is a reduction in the amount of expected single-copy sequences. Why? If the 900 nt long fragment contains both single copy and repetitive sequences it will reanneal under these conditions as a repetitive sequence, and thus the amount of DNA that appears in the single-copy fraction will be reduced proportionally. Therefore, that single-copy sequence is interspersed with repetitive sequences. If for example the single copy fraction is determined to account for 40per cent of the genome when 300 nt fragments are used, and we calculate that the genome has 30per cent single-copy sequences when a 900 nt tracer is used, 10per cent (40per cent-30per cent) of the 900 nt fragments contain single-copy and repeat fragments and that 25per cent (10per cent/40per cent) of the single-copy DNA is interspersed with repetitive sequences at an interval of 300-900 nt.

A final concern is the size of the repeat units. This can be obtained by using a long tracer (5000 nt, for example) and reannealing to an appropriate Cot such that only repeat sequences bind to the tracer. The product is then treated with an enzyme that cuts only single-stranded DNA. This enzyme is S1 nuclease. After digestion you obtain products that only contain repeat sequences. These products are then sized to give the average, mode and range of repeat sequences.

REPEATED SEQUENCES

The intermediate and fast components are composed of sequences that are found many times in the genome. These sequences are called repetitive sequences and can vary in size from a 100 bp to 1000 bp or more. Furthermore, these sequences have undergone sequence divergence by the addition or deletion of sequences or by changes in the base pair sequence. Thus, the repeated sequences themselves show some divergence. An example of a highly repetitive sequence is the repeat found to be associated with the knob heterochromatin of corn. It ranges from 3-5 $\times$ 10^5 copies on a small knob to 1 $\times$ 10^6 on the large knobs. This sequence is unique to knobs and is not found associated with any other heterochromatic regions of corn. An example of a functional repeated sequence in plants is the corn storage proteins, zeins. Two major classes of zeins exist, the 22 and 19 kd classes. Sequence analysis has shown that both classes have the same structure.

The only difference between the 22 and 19 kd class is the repeat unit. The 22 kd class has 8 repeat units and the 19 kd class has 7 repeat units. Estimates have been made of the number of copies of these genes and 30-50 copies of the 22 kd class are found in the corn genome. Thus, the repeat unit would be represented 240-400 times in the genome.

EVOLUTION OF REPEATED SEQUENCES IN CEREALS

The analysis of repeated sequences has also provided experimental evidence which supports the current model of cereal speciation. The following tables and the associated diagram detail the results. These show that once a repeat enters the lineage it remains. Therefore group I repeats are found in all of the species because it is the most ancient repeat. Secondly, new repeats are added to the lineage and appear in all species which subsequently diverged from the lineage. Finally, species-specific repeats are formed in each species after that species has diverged from the common lineage. The distribution of the lineage specific repeat supports the speciation model.

ESTIMATING THE NUMBER OF EXPRESSED GENES

Reassociation kinetics can also estimate the number and abundance of expressed genes. These experiments are performed using high concentrations of RNA and tracer levels of either DNA or cDNA. These experiments are analagous to DNA reassociation experiments except the units are expressed as moles of ribonucleotides per Litre per second and the value is called a Rot. If tracer amounts of DNA are allowed to reanneal only a portion of that DNA will form a duplex because not all the DNA will be expressed in the RNA population to which it is being hybridized. Let's use the example in the Genes V to derive an estimate for the number of genes that are being expressed. In this experiment 1.35per cent of the DNA hybridized to the RNA. Since RNA is single stranded the other anti-sense strand of the DNA would not have a partner with which it could hybridize. Thus actually 2.7per cent of the DNA is represented in this RNA population. If the genome size is 8.1 X 108 bp and the single-copy sequences represent 75per cent of the genome then we can estimate the complexity of the expressed DNA. $0.027 \times 0.75 \times (8.1 \times 108 \text{ bp}) = 1.7 \times 107$ bp If each gene is about 2000 base pair then the number of genes that is expressed is:

$$7 \times 107 \text{ bp} / 2000 = 8500 \text{ genes}$$

MITOCHONDRIAL GENOME ORGANIZATION

In comparison to the chloroplast genome, the size of the mitochondrial genome is quite variable.

Further, in comparison to the mitochondrial genomes of other species the size is quite large and variable. For example, animal mitochondrial genomes range in size form 15-18 kb, and fungi mitochondrial genomes range form 18-78 kb.

Species	Size (kb)
Oenothera	195
Turnip	218
Corn	570
Muskmelon	2400

Plants may code for more proteins than with species. For example, genes for ribosomes, subunits I and II of cytochrome oxidase and ATPase subunits are located on the mitochondrial genomes of plants.

When DNA from corn mitochondria was investigated with EM, several circular molecules of different sizes were detected. Once the genome was mapped it became apparent that a mechanism existed to generated these circles of different sizes. It is now understood how these molecules arise. First, lets look at the simple situation of turnip. Two direct repeats undergo intramolecular recombination to give the two smaller molecules:

$$218 \text{ kb} \rightarrow 135 \text{ kb} + 83 \text{ kb}$$

The mitochondrial genome of corn undergoes the same type of recombination, but the events are more complex. First, the master circles can be subdivided into two major subgroups:

$$570 \text{ kb} \rightarrow 488 \text{ kb} + 82 \text{ kb}$$
$$570 \text{ kb} \rightarrow 503 \text{ kb} + 67 \text{ kb}$$

Promiscuous DNA

Stern and Lonsdale (1982) hybridized mtRNA to a SstII digest of maize mt DNA and found that it hybridized to fragments known not to contain mt rRNA genes. The question of interest was - what was it hybridizing to? They next looked at a cosmid clone of corn mtDNA that hybridized to the mt RNA and found that it hybridized to a RNA molecule of the size of the cp 16S RNA gene. How could this have happened?

They next mapped the clone and compared it to the map of the corn cpDNA and found that the clone map was almost congruent with that of the the cpDNA 16S RNA region. This mapping showed that the two maps were nearly identical over a 12 kb region of DNA. These results suggest that cpDNA had been transferred to the mitochondrial genome. The observation that organelle DNA was found in other DNA compartments of the cell was extended by other researcher. Stern and Palmer looked at corn, mung bean, spinach and pea and found extensive evidence of cpDNA/mtDNA homology. These observations were extended to other DNA locations in the plant cell. Kemble *et al* (1983) demonstrated that mitochondrial DNA sequences are located in the nucleus of corn. Scott and Timmis (1984) showed that cpDNA sequences are found in the nuclear DNA.

CENTRAL DOGMA OF MOLECULAR GENETICS

The Central Dogma of Molecular Genetics states that the information that is found in DNA is used to produce mRNA molecules that are instrumental in the production of proteins. Therefore, the information flows directly from DNA to protein, via the RNA intermediate molecule. Recently it has been discovered that the information that is contained in the DNA is not always found in the RNA products used to make proteins.

It has now been demonstrated that mitochondria and chloroplast contain the biochemcial machinery to alter the sequence of the final transcription product. This process is called RNA editing. This process was identified in the following manner. Sequence analysis of a number of cytochrome c oxidase subunit II genes from non-plant species revealed that a tryptophan residue was invariant at several locations in the final protein product. But sequence analysis of this gene in several plant species revealed arginine at those positions.

Since a single base pair change in the codons for the two amino acids could generate this change (CGG for UGG), it was suggested that CGG encoded for tryptophan and not arginine in plant mitochondria. (This is the only change in codon usage that has been suggested for plants and has been postulated for several other genes as well.) But this change in codon usage was not universal, that is some CGG codons actually specified arginine in the final protein product. Furthermore, no amino acyl tRNA that recognized CGG was found to be charged with tryptophan, a prerequisite if this codon specification was actually real.

The solution to this dilemma was found by sequencing the mRNA products for cytochrome oxidase subunit II genes. It was found that in the mRNA the cytosine residue had been changed (edited) to uridine at the sequence location where the invariant tryptophan residue is found.

This changed the codon at that location to UGG which is recognized by a tRNA that carries the amino acid tryptophan. An analysis of three other plant mitochondrial genes where the same altered codon usage was predicted suggested that mRNA editing was also occurring at the codon and that a cytosine residue was edited to uridine. This editing process has also been detected in protozoa and it remains to be determined if RNA editing is a widespread function in mitochondria. A final point that this editing function highlights is that the sequence that is found in the DNA is not entirely and faithfully represented in the final protein product.

GENE EXPRESSION PROFILING

A second potential application of DNA microarray technology in safety evaluations concerns screening for biological effects of (fractions of) plant compounds that have been found by other profiling techniques to be present or modified in the genetically modified crop plant when compared to the parental line and other references. At RIKILT-DLO we are testing the feasibility of this strategy by analysing defined food compounds for possible effects on human intestinal gene expression. To this end, human intestinal cell lines are exposed for different time periods to various amounts of crop plant components and the mRNA isolated. The labelled mRNA is hybridized to DNA microarrays containing unknown intestine-specific genes, as well as already identified control genes known to be intestine specific or to be functional indicators (for toxicological processes, apoptosis, etc.).

As a source for unknown, intestine-specific genes we use cDNA libraries from human intestine biopsies and human intestinal cell lines which have been enriched by a subtractive cloning approach. Having established the usefulness of this approach for the identification of biological functions of defined plant compounds, these cDNA microarrays will be exploited to screen, for example, GM crop plants for possible effects on human health. The concept of substantial equivalence is broadly accepted as a basis for the risk assessment of GM crops and derived novel foods.

Data requirements for establishing substantial equivalence should be based on identifying unintended metabolic perturbations in the engineered crops that have been caused by genetic modification. A generic approach cannot be envisaged, however. In all cases the collation of data should be directed by the type of genetic modification and related consequences. It is recommended that unintended effects should be identified on a case-by-case basis, since such effects are dependent on the genotypic and phenotypic parameters of the new gene products involved.

The application of innovations in molecular genetics research will help to define the conditions under which the new food products can be marketed. The present approach has generated a sound scientific basis for the evaluation of the potential of unintended effects in GM crops and derived food products. The results may assist regulators and legislators to test and, if necessary, to improve the applicable regulations.

Moreover, the technologies described and the results obtained may serve as a general framework for the risk assessment of GM crop plants, and may contribute to a better understanding by the public of recombinant DNA techniques in plant breeding and their implications. It is concluded that a parallel approach focusing on a comparative analysis using informative profiles of molecules at various integration levels (*e.g.* mRNA, protein, metabolites) including a toxicological profiling of relevant plant matrices, offers good prospects for the identification of hazards related to unintended effects. The choice of comparators and of external parameters to support a claim that a GM crop plant presents no unintended effects compared with its traditional counterpart should be based on sound scientific judgement. It is recommended that data are generated on the unprocessed part of the plant, such as the tomato fruit or potato tuber, used for human and/or animal consumption.

Furthermore, data banks should be set up containing information on natural variations of essential plant or other product constituents (*i.e.* crop mean values). The chemical fingerprinting technique, using a combination of off-line liquid chromatography and proton-NMR imaging, appears to be a powerful screening method for the detection of secondary effects in the metabolome that may be applied on a routine basis.

The carbohydrate profiling technique is useful for the detection of unexpected changes at N-glycan levels; in particular it is applicable to the detection of post-translational modifications in newly expressed proteins as well as in the whole GM crop plant. DNA microarray technology is rapidly evolving and much effort is being put into improving spot density, reduced production time, and increased reproducibility and sensitivity.

Improving the latter is critical because it makes it possible to use smaller amounts of starting material. It will depend on technical aspects such as the quality of the scanners and spotting machines, the development of fluorescent dyes with improved characteristics (*e.g.* narrow excitation and emission peaks; high level of photon emission; resistance to photo-bleaching), supports with reduced background and more target sequence binding capacity.

DNA microarrays generate a huge amount of complex hybridization data and major challenges are to develop more advanced computer software to help to find statistically significant correlations within, and between different experiments, and to link the data to sequence information and submitted expression profiles available in public databases. Application of these technologies in combination with other measurements of, for instance, the performance and quality may limit or even replace animal feeding studies aimed at the detection of unintended effects.

Together, these techniques, known as functional genomics, will enable us to answer questions about what happens to expression and protein composition when a plant cell changes metabolic state, for example, due to genetic modification or environmental factors. It is foreseen that this holistic view on the cellular machinery will be achieved soon, as the coincidence of genome sequencing with improvements in the analysis of expressed proteins is of growing importance and continues to be successful.

Eventually, microarrays for genomic studies will probably also be used to evaluate the numerous constituents of GM crop plants. Some constituents might not cause obvious changes in cellular behaviour or morphologic appearance but could cause subtle metabolic alterations that would show up when the mRNA content was interrogated by an array. It is likely that microarrays might be used to evaluate the life cycle of a plant much more precisely and to understand the complex metabolic control systems of plants. The ultimate test for GM crop plants will be acceptance by the public in large. It is therefore extremely important that industry, the scientific community and the regulatory authorities are able to respond to queries from the public by providing transparent information on the criteria for safety assessment. Improved safety assessment at the molecular level should refine and complement strategies by anticipating the potential risks and sources of unintended effects in GM plants.

GENETICAL MODIFIED ORGANISM IN DERIVED FOODS

The concept of substantial equivalence was originally developed through discussions at the Organisation for Economic Co-operation and Development, though, to a large extent, these discussions built on previous work done by the World Health Organization (WHO) and the Food and Agriculture Organization. In 1991, a Joint FAO/WHO Consultation had concluded that the evaluation of a food derived through modern biotechnology should consider both food safety and nutritional value using similar conventional food products as a standard and taking into account the processing of the food and its intended use.

In the 1980s, the OECD established a Group of National Experts on Safety in Biotechnology (GNE), which continued to work through to 1993. As an intergovernmental organization, OECD's GNE was attended by delegates nominated by the governments of the OECD member countries, who were, for the most part, representatives of those agencies and ministries with a responsibility for safety in biotechnology. As a result of the work of the GNE, the OECD published a number of important documents relevant to the safety assessment of biotechnology-derived products during this period, which deal with a range of issues, including safety considerations for industrial, agricultural as well as environmental applications of organisms derived by recombinant DNA techniques.

It was recognized at this stage that the safety assessment of an organism derived through recombinant DNA techniques would rely heavily on the knowledge of its parental organism, as well as on an analysis of how the new organism appears to differ from the parent. OECD recommended that the considerable data on environmental and human health effects of living organisms that exists should be used to guide the risk assessment.

By the early 1990s, there had already been large numbers of small-scale field trials of new crop varieties derived through recombinant DNA techniques, and it became clear that new foods derived from these varieties would be marketed during the 1990s. In order to proactively address food safety issues related to these novel foods, in 1990, the GNE established a Working Group on

Food Safety and Biotechnology comprising experts, mainly from the ministries and agencies responsible for food safety issues in OECD member countries. The main objective of the Working Group was to elabourate scientific principles and concepts to be used when evaluating the safety of new foods and food components of terrestrial microbial, plant or animal origin.

SAFETY ASSESSMENT OF FOOD ADDITIVES

The Working Group did not consider the safety assessment of food additives, contaminants, processing aids or packaging materials. Nor did it consider environmental safety issues which had been (or were being) addressed by other groups of the GNE. Early in the life of the Working Group, an important recognition was that traditionally, the safety of food for human consumption had been based on the reasonable certainty that no harm will result from intended uses under the anticipated conditions of consumption. Foods prepared and used in traditional ways have usually been considered safe on the basis of long-term experience, even though they may have contained natural toxicants or anti-nutritional substances.

Normally, new varieties of foods or crops have not been subjected to traditional toxicological testing. In fact, where toxicological testing has been applied to whole foods, the results have often been difficult to interpret. Although the OECD's Working Group recognized that modern biotechnology might extend the scope of genetic changes that can be made - and might even broaden the range of the possible sources of foods - it was recognized that this would not inherently lead to foods that are less safe than those developed through conventional techniques.

The Working Group therefore indicated that the evaluation of foods or food components derived through modern biotechnology does not require a fundamental change in established principles, nor does it require a different standard of safety. Realizing the importance of using examples of new foods to identify and demonstrate the applicability of the proposed scientific principles, the Working Group organized a number of meetings and intergovernmental consultations which included case study presentations of novel foods such as enzymes, genetically modified bakers' yeast, mycoprotein and a number of genetically modified varieties of crop species.

Although these case studies were not intended to be formal safety evaluations they were crucial in illustrating important concepts in the safety assessment of novel foods. Therefore, it was through their application in real examples that the concepts and principles were identified. The Working Group proposed a scientific approach to the evaluation of foods derived through modern biotechnology, which is based on a comparison with traditional foods that have a safe history of use. One of the main concepts developed was that of substantial equivalence, and in describing this concept, the Group stressed that this was a principle that had been used in the past (perhaps intuitively) even if it had not been articulated as such. The concept of substantial equivalence is described as embodying the idea that existing organisms used as food, or as a source of food, can be used as the basis for comparison when assessing the safety of human consumption of a food or food component that has been modified or is new. As previously indicated, in elabourating the concept of substantial equivalence, the OECD Working Group noted that food safety is considered as a reasonable certainty, that no harm will result from intended uses under the anticipated conditions of consumption and that the most practical approach to the determination of safety is to consider whether a genetically modified organism (GMO)-derived food is comparable to an analogous conventional food product.

The concept of substantial equivalence therefore relies on the existing history of safe food use of a conventional food product as a useful consideration in framing the safety assessment of a GMO-derived food by permitting the identification of similarities and differences which can be considered in the assessment. In 1996, participants at an expert FAO/WHO consultation recommended that safety assessment based upon the concept of substantial equivalence be applied in establishing the safety of foods and food components derived from genetically modified organisms.

Assessing substantial equivalence was recognized as not being a safety assessment per se, but a process which establishes that the characteristics and composition of the new GMO-derived food are comparable to those of a familiar, conventional food which has a history of safe consumption. A Joint FAO/WHO Expert Consultation on Foods Derived from Biotechnology was convened in 2000 to address food safety and nutritional questions regarding foods derived from GM plants, including a review of the scientific basis, application and limitations of the concept of substantial equivalence.

It concluded, based on the current application of substantial equivalence and alternative strategies that this concept contributed to a robust safety assessment framework. Moreover, the Consultation noted that substantial equivalence is a concept used to identify similarities and differences between GM food and a comparator with a history of safe food use which subsequently guides the safety assessment process.

While not a safety assessment per se, the substantial equivalence approach allows the structuring of the safety assessment through characterizing the similarities and identifying the differences which can then be the focus of further consideration. This approach has been referred to as a useful safety standard. It therefore permits inference that the new food under consideration will be no less safe than the conventional food under conditions of similar exposure, consumption patterns and processing practices. Substantial equivalence is therefore clearly not intended to be a measure of absolute safety, but instead recognizes that while demonstrating absolute safety is an impractical goal, demonstrating that there is reasonable assurance that the GMO-derived product under consideration is no less safe than a conventional food product is an achievable goal. Since it's development, the use of the substantial equivalence concept in the safety assessment of GMO-derived foods has been subject to misinterpretation and criticism.

The comparative nature of the substantial equivalence concept without prescribing the extent of the phenotypic and compositional comparisons has led some to criticize the concept as not being measurable, and therefore inappropriate in safety assessment. This and other criticisms relate, in part, to the mistaken perception that the determination of substantial equivalence was the end point of a safety assessment rather than the starting point.

After several OECD countries had gained experience with safety assessment of GMO-derived foods, an OECD workshop examined the effectiveness of the application of substantial equivalence in safety assessment. This workshop concluded that the substantial equivalence approach provides equal or increased assurance of the safety of foods derived from genetically modified plants, as compared with foods derived through conventional methods.

SAFETY OF NOVEL FOODS AND FEEDS

In 2000, the OECD Task Force for the Safety of Novel Foods and Feeds also reviewed the substantial equivalence concept, its interpretation and its application. While noting that the majority of international guidance documents addressing the safety assessment of genetically modified (GM) plants had interpreted substantial equivalence consistently, this Task Force reported that there were differences in how it was applied and that these differences needed to be resolved.

The Codex Ad Hoc Intergovernmental Task Force on Foods Derived from Biotechnology pursued the development of international guidance on the safety assessment of GMO-derived foods. To date, this Task Force has developed proposed draft principles and guidelines for the safety assessment of foods derived from modern biotechnology. These guidelines interpreted the concept of substantial equivalence as a way of structuring the safety assessment, consistent with the FAO/WHO interpretation.

The 'Draft Principles for the Risk Analysis of Foods Derived from Modern Biotechnology' and 'Draft Guideline for the Conduct of Food Safety Assessment of Foods Derived from Recombinant-DNA Plants' have been forwarded by the Codex Task Force to the 25th Session of the Codex Alimentarius Commission for adoption at Step 8 of the Codex procedure. A proposed draft guideline for the food safety assessment of foods produced using recombinant-DNA organisms continues to be developed. Substantial equivalence: the application A conclusion from a 1995 WHO workshop on 'Application of the Principles of

Substantial Equivalence to the Safety Evaluation of Foods or Food Components from Plants Derived by Modern Biotechnology' summarized the relationship between substantial equivalence and safety assessment as follows: Establishment of substantial equivalence is not a traditional safety assessment in itself, but a dynamic, analytical exercise in the assessment of the relative safety of a new food or food component to an existing food or food component. The application of the substantial equivalence concept is a key step in structuring the safety assessment of GMO-derived foods. Through appropriate analysis, the new food is compared phenotypically and compositionally to its conventional counterpart with a history of safe use. In this compositional comparison, the characteristics, including levels of key nutrients and toxicants, are considered relative to those of the conventional counterpart taking into account the natural variation for such characteristics.

When a genetic modification results in the insertion of a specific trait, the comparative analysis of the new food will identify both the intended effect (inserted trait) and the potential unintended effects of the modification. Further assessment can then focus on new or altered characteristics for which no history of safe use can be established.

One of the important benefits of applying the substantial equivalence concept is that it provides flexibility which can be a powerful tool in terms of food safety assessment. The comparative approach to structuring the safety assessment can be applied at several potential levels along the food continuum (*i.e.* harvested primary food material or unprocessed food product, individual processed fractions, or final food product or ingredient).

While from a practical point of view, the compositional comparison should typically be applied at the level of the unprocessed food product, the flexibility of the concept permits the determination to be targeted to the most appropriate level based upon the nature of the product under consideration. Where multiple fractions from a single source are destined to different food products, the comparative approach might be targeted at the level of the unprocessed food product in order to permit the safety assessment to apply to all derived fractions (*e.g.* for soybean, where multiple fractions are used widely in foods, assessment at the level of the seed is appropriate).

However, where a single fraction of a particular raw material is used as human food, then the comparison can focus at the level of the single fraction, thereby simplifying the safety assessment process (*e.g.* for canola, where the processed oil is the only fraction consumed by humans, safety assessment may appropriately be focused on comparison of the oil composition of the novel variety with traditional canola oil composition).

Application of the substantial equivalence concept in the safety assessment of a GMO-derived food depends on the identification of an appropriate comparator with an acceptable history of safe food use. It also requires that sufficient analytical data be available in the literature or be generated through analysis to permit an effective comparison. These requirements present a key limitation of the substantial equivalence concept since the consideration of similarities can only provide assurance of safety relative to those components assessed for the particular comparator. The choice of the comparator is therefore crucial to the effective application of substantial equivalence in establishing the safety of a GMO-derived food.

An appropriate comparator must have a well-documented history of use. If adverse effects have been associated with the particular food type, specific components of the food which are considered to be causative of those adverse effects should be described and well characterized in order to permit effective comparison. A joint FAO/WHO Expert Consultation on Biotechnology and Food Safety considered the application of substantial equivalence in the safety assessment of GMO-derived foods.

The consultation recommended that applicacation of the substantial equivalence concept entail consideration of the molecular characterization of the new food source; phenotypic characterization of the new food source in comparison to an appropriate comparator already in the food supply; and the compositional analysis of the new food source or the specific food product in comparison to the selected comparator. The guidance further elabourates on compositional comparison by highlighting what information should provide sufficient information to permit effective comparison.

The recommended focus of the compositional comparison is on analytical comparison of those components identified in the food source in question which are nutrients which provide a substantial impact in the overall diets (key nutrients) and toxicologically significant compounds known to be inherently present in the species (key toxicants). In addition, there is recognition that additional components might be identified for analysis based upon the molecular and phenotypic characterization and the nature of the genetic modification. In addition to key nutrients and toxicants, additional parameters may be appropriate for assessing the potential for unintended effects of a genetic modification.

When modifications are directed at metabolic pathways of key macro or micro nutrients, the possibility of an impact on nutritional value is increased, particularly if that food is a major dietary source of the nutrient affected. The potential for unintended effects would be determined, in part, by the nature of the intended alteration (*e.g.* consideration of fatty acid profile if an enzyme involved in fatty acid metabolism is introduced) and the data from molecular and phenotypic characterization.

The consideration of key nutrients and key toxicants in the comparison which is essential to applying a substantial equivalence in safety assessments also introduces another limitation of the concept. The nature of the comparative approach with respect to nutrients limits its universality since the relevance of nutrients in a particular crop are dependent on consumption patterns which might vary from region to region. Where differences in consumption exist for a particular crop, these must be considered in the identification of the key nutrients for assessment.

Particularly True for Crops

This is particularly true for crops which form a significant portion in the diet in a particular region. The comparative approach to assessment can be applied in each region, but conclusions for one region will not automatically hold for another region if there are significant differences in consumption patterns and processing practices. Another potential regional limitation is related to the application of the concept of substantial equivalence as opposed to an inherent limitation of the concept per se.

It may be difficult, particularly in developing countries, to apply the concept to assess the safety of foods where adequate nutritional databases are not available for a given population. A safety assessment using the substantial equivalence approach does not demonstrate that a GMO-derived product is identical to its conventional comparator since the compositional comparison does not take into account all components.

However, application of the guidance provides assurance that the comparison has considered those components most likely to be relevant to the safety of the product as it is expected to be consumed in a particular region. Recognizing the importance of there being compositional data available for applying substantial equivalence, the OECD Task Force for the Safety Assessment of Novel Foods and Feeds has focused on the development of science-based consensus documents containing information on the nutrients, anti-nutrients or toxicants, product use and other data relevant to the assessment.

The OECD have published such documents for potatoes, sugar beet, soybean and low erucic acid rapeseed (canola). Such guidance will add to the understanding of appropriate parameters for a comparative assessment of composition while recognizing that additional parameters may be relevant to the safety assessment dependent on differences in consumption pattern. The international guidance developed by the OECD and FAO/WHO has been practically applied to the safety assessment of GMO-derived food products in several countries. In order to facilitate such assessments, specific guidance documents which embrace the substantial equivalence concept have been published. The majorities of these currently address the safety assessment of genetically modified plants and have consistently interpreted the concept of substantial equivalence. Internationally, these guidance documents have been applied to the assessment of a significant number of GMO-derived plant products over a period of more than eight years, demonstrating that the concept of substantial equivalence can be applied effectively in the safety assessment of novel foods.

CONCEPT OF SUBSTANTIAL EQUIVALENCE

The comparative approach and hence the concept of substantial equivalence has been applied to the safety assessment of new and modified foods derived from plants developed using traditional breeding practices over the last 20 years and more recently to those products derived from recombinant DNA technology.

Just as the OECD Working Group utilized case studies, the application of substantial equivalence in structuring the safety assessment of novel foods can be illustrated through discussion of the modifications made to rapeseed oil through traditional and recombinant DNA techniques.

Low-erucic Acid Rapeseed Oil

In 1987 low-erucic acid rapeseed oil (LEAR oil) was given generally recognized as safe (GRAS) status in the United States. Although this product was not produced as a result of recombinant DNA technology, it was considered to be a 'novel food'. The LEAR oil case study illustrated the application of the principles developed by the Working Group (established by the GNE) for assessing the safety of foods by applying the concept of substantial equivalence. The novel trait of LEAR oil was the low erucic acid content when compared with traditional rapeseed oil. In this case precise comparisons between LEAR oil and other vegetable oils could not be made because vegetable oils vary in composition depending upon the variety of plant and the growing conditions.

The concept of substantial equivalence was applied to assess the safety of LEAR oil by comparing the individual fatty acid components to similar components present in other traditional oils including soy, corn, peanut, safflower, olive and sunflower. Except for the low levels of erucic acid, the individual fatty acid components of the LEAR oil were comparable to those similar fatty acids found in common vegetable oils.

Dietary exposure estimates for average and upper limit intakes for LEAR oil used by itself and as a component of blended oil products including shortening, margarine, salad oil and vegetable oil did not raise any safety concerns. The data considered in assessing the nutritional adequacy and digestibility of LEAR oil were similar to those that would be considered for any new oil (*i.e.* human and animal feeding studies).

Due to concerns regarding the safety of the erucic acid component in both traditional rapeseed and LEAR oils, toxicological studies were an important consideration in the safety assessment. These studies included a review of an extensive toxicological database and a scientific rationale supported by the results of animal feeding studies in several species with a range of vegetable oils. The development of LEAR/canola and the application of the comparative approach to directing its assessment is illustrative. We can further build on this illustration by considering the application of recombinant DNA technology to canola to develop herbicide-tolerant varieties.

In these cases, considerations for the safety of the oil derived from these varieties follow the same approach. In this case the appropriate comparator is now the unmodified canola variety. The database used in establishing the safety of LEAR/canola oil provides an effective tool for the comparison of the composition of the oil derived from the herbicide-tolerant variety under consideration.

Comparison of the oil from new canola varieties developed through recombinant DNA technology with the oil from unmodified canola with the same food use is an effective approach to determine its safety. Consideration is appropriately given to key nutrients (*i.e.* fatty acids) and key toxicants (*i.e.* ensuring that the level of erucic acid is sufficiently low for safe consumption). The comparative approach can also be applied in the safety assessment of other compounds in plants such as canola.

The canola meal, which may be used as an animal feed, would require a separate evaluation since measurable amounts of the introduced protein may be present in the meal, unlike in the refined oil. In the case of the meal, the safety of that protein in animal feeding may need to be demonstrated if the protein is new to the feed supply.If, based on the compositional comparison of the canola meal to an unmodified counterpart, the only difference is the presence of the introduced protein, then the next steps in the safety assessment would focus on determining if that new protein had the potential to be toxic or impact the nutritional quality for feed use.

The successful lowering of erucic acid led to continued interest in the compositional modification of canola oil. For example, plant breeders have used mutagenesis to genetically alter the plant's fatty acid biosynthetic pathways to obtain specialized fatty acid compositions. Canola oil has been developed with the linolenic acid content reduced from approximately 10per cent to less than 3per cent. Although high levels of linolenic acid are desirable from a nutritional point of view, they are undesirable in terms of chemical stability.

Other recent developments in canola oil compositional changes include the application of mutagenesis to produce high levels of oleic acid. The resulting high oleic acid-producing cultivar was then crossed to low linolenic cultivars to create a high oleic/low linolenic line. Intentional modifications in the composition of canola oil may lead to the production of canola varieties that are not like other commercial varieties. For example, in the last 10 years the application of recombinant DNA technology has resulted in the production of increased levels of lauric and myristic acids in canola oil. Similarly, the application of mutagenesis has produced higher levels of oleic acid in canola oil. For these canola varieties the fatty acid profiles and levels will not fall within the ranges defined in the Codex standard for Edible Low Erucic Acid Rapeseed Oil or the Codex draft standard for Named Vegetable Oils (which includes low erucic acid rapeseed oil).

In cases where the fatty acid composition of canola oil has been intentionally modified so that commercial canola varieties cannot be used as a comparator, the fatty acid component(s) will need to be considered on an individual basis using other commonly consumed oils as appropriate for comparison where such fatty acid component(s) are present in similar levels.

Substantial equivalence could be applied at the component level to assess the safety of the oil produced since these fatty acids have a safe history of consumption as a significant component of other edible oils. In addition to fatty acid profiles and levels, modifications may also result in alterations in the chemical structure (*e.g.* saturation, chain length and triglyceride structure) that may have nutritional consequences or result in changes in digestibility.

Chemically altered fatty acids will need to be evaluated on their merit, which may involve a combination of nutritional and toxicological in vivo and in vitro testing. Laurate canola provides an example of oil that is not compositionally identical to any other food oils, although it shares many similar characteristics.

Unlike other commercial canola varieties which contain no detectable lauric acid, laurate canola produces high levels of lauric acid. Laurate canola also produces lower levels of oleic acid and higher levels of myristate compared with the Codex specifications for those fatty acids in low erucic acid rapeseed oil. Other fatty acids in laurate canola, including palmitic, palmitoleic, stearic, linoleic, linolenic, gadoleic, eicosadienoic, behenic and lignoceric, are similar to those levels found in commercial canola varieties.

Levels of the naturally occurring toxicant erucic acid are very low. Substitution of laurate canola for coconut and palm kernel oils does not raise any safety concerns for the intended uses because the major components, laurate and myristate, are identical. The safety assessment of novel foods based upon the concept of substantial equivalence relies on comparison with the conventional foods with a long history of safe use. The comparator can be the species itself, a product derived from that food source or a similar component from a different species (*i.e.* fatty acids in vegetable oils). The food safety issues for organisms that have been genetically modified are of the same nature as those that may occur through other ways of genetic modification, such as traditional breeding.

These include potential food safety concerns such as toxicity or allergenicity. Once the safety assessment is completed it is reasonable to assume that the novel food does not pose a risk different to those of traditional foods that have been safely part of the diet for many years. The term 'substantial equivalence' and its application in safety assessment has been criticized or challenged as lacking a clear definition and therefore being imprecise. However, when considered as a concept, the lack of a measurable definition does not restrict its application. Stated most simply, substantial equivalence encourages investigators to compare a product which they have to assess with one with which they are already familiar. The application of this concept to the assessment of foods derived from GMOs is therefore intended to permit the creation of a linkage between the new GMO-derived food and a familiar, conventional food in terms of their characteristics and composition. Over the past eight years, the concept of substantial equivalence has been consistently interpreted and practically applied by numerous countries assessing the safety of foods derived from genetic modification.

It has been demonstrated that the concept of substantial equivalence can be applied effectively in the safety assessment of those genetically modified foods developed for commercialization to date. It is noteworthy that the Codex Ad Hoc Intergovernmental Task Force on Foods Derived from Biotechnology has supported the comparative approach as a key element in structuring the safety assessment of GMO-derived foods, signifying that there is significant international consensus emerging on the application of substantial equivalence.

While not yet a formal consensus, the agreement of the Codex Task Force to forward the 'Draft Principles for the Risk Analysis of Foods Derived from Modern Biotechnology' and 'Draft Guideline for the Conduct of Food safety Assessment of Foods Derived from Recombinant-DNA Plants' to the Codex Alimentarious Commission for adoption signifies emerging consensus.

Developments in Rapeseed Oil (canola)

Rapeseed breeding in Canada began soon after the crop was introduced during World War II. The initial goals of breeding were directed towards improving agronomic characteristics and oil content of rapeseed. Nutritional experiments conducted as early as 1949 indicated that consumption of large amounts of rapeseed oil with high levels of erucic acid could be detrimental to experimental animals.

Concerns about the nutritional safety of rapeseed oil and the potential impact on human health stimulated plant breeders to search for genetically controlled low levels of erucic acid in rapeseed oil. After 10 years of backcrossing and selection to transfer the low erucic acid trait into agronomically adapted cultivars, the first low erucic acid varieties, Brassica napus and B. campestris were released in 1968 and 1971, respectively. Rapeseed meal is used exclusively in Canada as a high-protein feed supplement for livestock and poultry.

Prior to the late 1970s, the use of this oilseed processing byproduct as an animal feed was limited by the presence of glucosinolates in the seed. The low palatability and the adverse effect of glucosinolates due to their anti-thyroid activity led to the development of varieties of rapeseed which have combined low levels of both glucosinolates and erucic acid (also known as 'double low' varieties). Canola breeding programmes in the 1980s and 1990s have produced cultivars with higher yields, increased oil and protein contents, earlier maturity, yellow seeds, reduced green seed and improved disease, insect and herbicide resistance.

TRANSFER OF DNA FROM GMOS IN FOODSTUFFS

There is potential for the presence of live GMOs in foodstuffs such as yoghurt and fermented sausage produced with starter cultures containing genetically modified organisms or perishable foods treated with protective cultures of GM organisms. The environment within living food may permit gene transfer by conjugation or transduction and this has been seen at high frequency in complex food matrices. Free recombinant DNA may also be present in food products resulting from the killing of GMMs during food processing. The human gut microflora is extremely complex, being made up of more than 500 different species of bacteria and climax microbial populations reaching up to 1011 CFU/ g luminal contents in the colon. The high density and species diversity of bacteria present in the colon provides ideal conditions for the transfer of DNA between bacteria.

However, the diversity of the human gut microflora also presents significant problems in monitoring the DNA transfer events between bacteria and, in particular, monitoring the fate of recombinant DNA, which may be ingested in food. It has been determined that only a fraction of the bacteria present in the human gut microflora may be cultured under labouratory conditions and characterized.

Many species found in the human gut microflora defy culture by standard microbiological techniques, and selective agars and growth conditions exist for only a small proportion of the predominant species present. Much progress as been made in recent years in the development of culture-independent methods for identifying bacteria present in natural environments.

Using the phylogenetic information present in 16S rRNA and fluorescent in situ hybridization techniques, it is now possible to identify bacteria in situ in natural environmental samples including human faeces. However, such techniques tell us little about the genetic capabilities of bacteria and by themselves do not allow us to monitor the distribution or indeed dissemination of specific genetic determinants amongst natural assemblages of bacteria.

Thus, studies on the transfer of DNA between members of the human gut microflora and from GMOs to members of the human gut microflora have employed specific marker systems borne on the mobile genetic elements or on recombinant DNA. Traditional marker systems used in pure culture microbiology such as the lac operon or X-gal are of little use when employed to monitor DNA transfer events in complex assemblages of gut microorganisms, because of the widespread occurrence of such genes amongst the mammalian gut microflora.

ANTIBIOTIC-RESISTANCE DETERMINANTS

Antibiotic-resistance determinants, when shown to be absent from background bacteria, have been widely employed in labouratory-based studies. However, such marker systems cannot be used in studies with human volunteers because of the emergence of antibiotic-resistance determinants in pathogenic strains. Novel marker systems such as green fluorescent protein are now being employed in labouratory studies to monitor the persistence of GMOs in the human gut microflora and DNA transfer events between bacteria. Another hurdle to monitoring the dissemination of recombinant DNA in complex microbial consortia and investigating the factors governing DNA transfer in natural environments is the non-expression of marker genes in novel bacterial hosts.

Marker genes transferred between bacteria may not always be expressed in the recipient cell. Thus not all DNA transfer events will be detected by existing marker systems. One way around this problem is to use a specific plasmid transfer system comprising a donor and recipient bacterium in which the marker gene is known to be expressed, and then examining the factors governing DNA transfer between these two strains in models of the ecosystem under investigation.

Recent advances in molecular biology offer some hope that more reliable means of monitoring the dissemination of genetic determinants amongst diverse bacteria may be developed. Fluorescent in situ hybridization is now being employed to enumerate bacteria in natural assemblages by hybridizing 16S rRNA targets to fluorescently labelled DNA probes and visualization with fluorescent microscopy. Similarly, it is now possible to visualize PCR products within bacterial cells fixed on microscope slides.

By combining these techniques it may soon be possible to visualize concomitantly, fluorescently labelled PCR products of single-copy genes and labelled 16S rRNA sequences within a single cell in environmental samples. To enable us to ask specific questions about the human gut microflora, in vitro and in vivo models of the human gut have been developed.

In vitro models generally employ anaerobic chemostats of varying degrees of complexity or short-term batch cultures of human faeces or luminal samples extracted from animals. They allow generation of data on the effect of specific biotic and abiotic environmental parameters on the gut microflora, which may otherwise be unavailable due to inaccessibility of the target environment in the gastrointestinal tract or for ethical reasons. Freter *et al.* studied the transfer of plasmid DNA between E. coli strains using a single-stage continuous flow culture model of the mouse gut microflora. This in vitro model of the mouse caecal microflora simulated microbial interactions observed in the mouse gut. Data generated from the in vitro model also allowed validation of mathematical models of plasmid transfer between E. coli strains in the presence of the complex mouse gut microflora. Rang *et al*. employed short-term batch cultures of luminal extracts from the mouse gastrointestinal tract to monitor transfer of plasmid RP1 between E. coli strains. More complex in vitro models of the human gastrointestinal tract developed recently have yet to be employed to monitoring DNA transfer events in the human gut microflora.

In vitro models allow us to examine interactions between specific members of the gut microflora or between the whole human faecal microflora and introduced GMOs and non-GMOs under the physiological conditions of the human gut.

They provide a vital first stage in the biosafety assessment of GMOs intended for use in human food. However, such models do not take into consideration the biological complexity of the mammalian gut ecosystem. In vitro models may not simulate the biological surfaces or the input of endogenous substrates and secretions of the human immune system, which may play a vital role both in the maintenance and establishment of the gut microflora, and the frequency of DNA transfer events.

Gnotobiotic and germ-free animals have been employed as in vivo models of the human gastrointestinal ecosystem for some years. Gnotobiotic technology provides a useful tool in studying the interactions between specific members of the human gut microflora, *e.g.* the role played by specific members of the microflora in colonization, resistance to pathogenic bacteria, bacterial pathogenesis, and metabolism of dietary constituents or xenobiotic compounds. Transfer of plasmid DNA between specific members of the human gut microflora has also been studied using di- or poly-associated animals. Conventional animals may not provide information of direct relevance to the human gut ecosystem because of the major differences in the composition of the gut microflora between animals and man. To circumvent these problems, germ-free animals have been associated with groups of bacteria from the human gut microflora or the whole human gut microflora.

Such human flora-associated (HFA) animals provide an essential tool in studies of human gut microbiology in that they reflect the complexity of the human gut ecosystem in terms of microbial species and numbers, and they provide some of the biotic parameters provided by the mammalian mucosa, immune system and digestive functions.

Such in vivo models provide an extremely useful tool in the generation of data on the safety of GMOs in food until a greater degree of confidence in the safety of such products is attained and human volunteer studies can ethically be considered.

PLASMID DNA FROM BACTERIA IN VIVO

Very little is known about the extent to which transformation and transduction contribute to DNA transfer between bacteria in the human gastrointestinal microflora. Only one study on bacterial transformation in fresh human saliva samples has been presented in the literature. No reports of transformation or transduction in lower regions of the human gastrointestinal tract were found after exhaustive search of the literature. This is probably due to the fact that conjugation has been viewed as the dominant contributor to DNA transfer between bacteria in natural environments, because of the ubiquitous occurrence of conjugative elements amongst diverse bacterial species. Thus most studies looking at DNA transfer in the gut have concentrated on conjugation.

Earlier, it was proposed that the frequency of plasmid transfer in a given environment will be a function of the interaction between environmental factors and the bacterial populations involved, characteristics of the plasmid itself and the mode of conjugation employed by the plasmid. Some of these parameters have been studied in vivo in animal models.

Effect of Antibiotics on Plasmid Transfer

A number of early investigators observed the transfer of plasmid DNA between bacteria in the gastrointestinal tract of human volunteers. However, many of these studies were conducted with antibiotic-resistance determinants in E. coli strains and in the presence of antibiotic selective pressure. In the light of the recent emergence of pathogenic bacteria with multiple antibiotic resistance determinants it is no longer considered ethically justifiable to carry out such investigations in human volunteers. Ingestion of antibiotics leads to perturbation of the natural homeostasis of the human gut microflora and presents a selective pressure for the evolution of antibiotic resistance strains through mutation and transfer of resistance determinants. The human gut microflora is commonly exposed to antibiotics, both through clinical practice and use of antibiotics in animal husbandry. Similarly, antibiotic-resistant bacteria have been isolated from a wide range of human foodstuffs.

Duval-Iflah *et al.* observed the transfer of an R plasmid from Serratia liquifaciens to an E. coli recipient strain, naturally present in the human microflora of the HFA mice. The donor S. liquifaciens strain became established in the HFA mice at between 106 and 107 CFU/g faeces. Transfer was observed 12 days after inoculation with the donor strain and their numbers ranged from between 103 and 105 CFU/g faeces. The number of transconjugants greatly increased after addition of selective antibiotic in drinking water. Morelli *et al.* showed that a subtherapeutic dose of selective antibiotic (erythromycin at 10 μg/ml drinking water) increased the level of pAMβ1 transfer from L. reuteri to Ent. faecalis from 1 × 10-7 to 1 × 10-4 transconjugants per donor in anexic mice (*i.e.* gnotobiotic mice associated with donor and recipient strains).Although selective antibiotic pressure previously has been employed to demonstrate transfer of genetic elements between bacteria in animal models and in humans, the observation that subtherapeutic doses of selective antibiotic increase the rate of plasmid transfer has important implications for both the design in GMOs and the use or misuse of antibiotics in clinical practice and agriculture.

RECOMBINANT DNA IN VIVO

The ability of GMOs, which may be ingested in a viable form in fermented foods, to persist in the human gastrointestinal tract will greatly determine the extent to which recombinant DNA may be transferred to the human gut microflora.Similarly the stability of recombinant DNA within GMOs in vivo is of major interest when designing GMOs for use in food or in the human gut itself. Stable maintenance of recombinant DNA will greatly determine the ability of a GMO to effectively carry out the task for which it was designed. On the other hand, a certain degree of recombinant DNA instability may limit the risks posed by recombinant DNA persisting in non-target ecosystems.

Gruzza *et al*. investigated the colonization potential and recombinant DNA stability of genetically modified L. lactis strains in germ-free mice. All L. lactis constructs reverted to plasmid-free derivatives in the digestive tract of anexic mice and plasmid-free derivatives became dominant over plasmid-bearing parental strains. Two high copy number, non-self transmissible plasmids exerted strong ecological disadvantage on plasmid-bearing L. lactis strains and were quickly lost from the digestive tracts of the anexic mice.

Transfer of the conjugative plasmid pIL205 was observed between an Ent. faecalis strain colonizing the digestive tract of anexic mice and an L. lactis donor strain. Both the L. lactis donor strain and Ent. faecalis (pIL205) transconjugants were lost from the anexic mice 10 days after dosing with the donor strain. Thus it appears that carriage of recombinant DNA on non-conjugative plasmids and indeed conjugative plasmid DNA, imposed considerable ecological disadvantages on plasmid-bearing strains in anexic animals. Schlundt *et al*. investigated transfer of pAMβ1 between L. lactis donor and recipient strains in gnotobiotic and conventional rats. In the conventional animals, both donor and recipient strain were quickly lost from the system and no transconjugants were recovered.

In gnotobiotic animals both donor and recipient strain colonized the gastrointestinal tract. Tranconjugants were recovered within a few days of dosing and maintained at between 103 and 105 CFU less than the recipient strain in faecal samples. Upon sampling of luminal contents of the jejunum, caecum and colon it was observed that carriage of the plasmid seemed to endow a competitive advantage to transconjugants in the small intestine but not in the caecum. The population of transconjugants was maintained at about 104 CFU/g throughout the intestine but numbers of recipient strain varied from between 104 and 105 CFU/g jejunum contents to 108-109 CFU/g caecal and colonic contents. Clearly the ability of plasmid DNA either to impose ecological advantage or disadvantage on the bacterial host will depend on the plasmid and bacterium involved, and may very well vary considerably between different environmental habitats or microhabitats.

Effect of Microbial Interactions on DNA Transfer

The human gut microflora exerts considerable colonization resistance on extraneous microorganisms entering the colonic environment. Limiting the persistence of a GMO in a given environment will also limit its ability to undergo DNA transfer with the microflora. Gruzza *et al.* investigated the ability of members of the human gut microflora to limit the colonization ability of and DNA transfer from, genetically modified Lactococcus lactis strains, similar to those likely to be used in the food industry.

L. lactis strains bearing the conjugative plasmid pIL205 and/or a non-conjugative but a pIL205 mobilizable plasmid pIL253 were used. Gnotobiotic mice were associated with four strictly anaerobic bacteria commonly found in the human gut microflora, *i.e.* Bacteroides sp., Bifidobacterium sp., Peptostreptococcus sp. and Ent. faecalis.

No transfer of plasmid DNA was observed from L. lactis donor strain to Bacteroides sp., Bifidobacterium sp. or the Peptostreptococcus sp. However, upon addition of Ent. faecalis to the polyassociated mice, Ent. faecalis transconjugants were observed despite the competitive exclusion imposed on the L. lactis donor strain by the colonizing Ent. faecalis strain. Gene transfer from the L. lactis strains was also investigated in HFA mice. HFA mice were dosed with L. lactis donor strain, bearing pIL205 and pIL253, seven days after introduction of the human faecal microflora.

From an initial inoculum of 8 × 108 CFU, no L. lactis donor cells were recovered from the faecal samples 24 hours after dosing (detection limit 102 CFU/g faeces). However, 21 hours after introduction of the donor strain, transconjugant facultatively anaerobic streptococci were recovered resistant to antibiotic markers borne on each of the two plasmids.

The number of transconjugants increased until day 15 after dosing. Thus, although rather quickly eliminated from the HFA mice, the L. lactis donor strain was able to transfer its plasmid DNA (both the conjugative pIL205 and the non-conjugative but mobilizable pIL253) to members of the human gut microflora in HFA mice.

Brockmann *et al.* investigated the transfer of plasmid pAMβ1 and pLMP1 (bearing a genetically modified proteinase gene) from L. lactis donor strains to a proteinase-deficient recipient strain in gnotobiotic and conventional rats.

No transfer of the proteinase gene to the recipient L. lactis strain was observed in anexic or conventional rats. Plasmid pAMβ1 was transferred between L. lactis donor and recipient strains in anexic animals but not in conventional rats where both donor and recipient were quickly lost from faecal samples. Transfer of pAMβ1 from the L. lactis donor strain to Ent. faecalis (or a close relative) was observed in one conventional rat.

Using an in vitro single stage continuous flow culture (CFC) of human faecal microflora and in vivo human flora-associated (HFA) rat model of the human colonic microflora we have shown that the human gut microflora exerts a considerable barrier to persistence of L. lactis strains in the colonic ecosystem. A plasmid-free L. lactis strain MG1614 was washed out from the in vitro model at a rate greater than expected for non-growing cells.

Similarly, L. lactis MG1614 introduced into HFA rats was eliminated from the HFA rats within eight days. By continuously dosing HFA rats with L. lactis MG1614 in the drinking water we succeeded in maintaining a stable population of L. lactis MG1614 in the rats. It was hoped that this strain implanted in the microflora of HFA rats would act as a recipient in plasmid transfer studies with a closely related L. lactis strain. Transconjugants were maintained at between 104 and 105 CFU/g faeces for 27 days. Thus despite the fact the human gut microflora exerts such a drastic barrier effect on the persistence of L. lactis strains in models of the human colonic microflora, transfer of plasmid DNA from an ingested L. lactis donor strain to Enterococcus spp. present in the microflora was observed.

Dietary Factors Affecting DNA Transfer

The ecological significance of the metabolic burden imposed on a bacterium by plasmid carriage will be greatly determined by the availability of substrates and nutrients to the cell. Similarly, densities of donor and recipient cells have been shown to affect conjugation frequencies in vitro. In our studies, transfer of plasmid pAMβ1 from an L. lactis donor strain to an Enterococcus spp. recipient strain naturally present in the microflora of HFA rats was found to be strongly affected by dose of donor strain employed.

Significant numbers of Enterococcus spp. (pAMβ1) transconjugants were recovered only at a donor cell concentration of 109 CFU/ml or above. The diet fed to the HFA rats also had an effect on transfer of pAMâ1 from the L. lactis donor strain to the Enterococcus spp. in vivo. A high-fat, low-fibre synthetic diet (similar to human high-fat diets) reduced the numbers of Enterococcus spp. (pAMβ1) transconjugants recovered from the faeces of HFA rats compared with the commercial rodent chow (higher levels of complex carbohydrate and fibre).

The presence of increased substrate availability in the caecum of HFA rats fed the rodent chow or increased surfaces provided by the higher fibre content of the rodent chow may be involved in the higher numbers of Enterococcus spp. (pAMβ1) transconjugants recovered from HFA rats fed this diet. Duval-

Iflah *et al.* looked at the effect of milk fermented with Lactobacillus bulgaricus and Streptococcus thermophilus on the transfer and mobilization of plasmids between E. coli strains in the gut of anexic mice.

Plasmid transfer studies were conducted in gnotobiotic mice associated with E. coli K12 donor strain and an E. coli recipient strain PG1 of human origin. The affect of milk fermented with L. bulgaricus, S. thermophilus or with both lactic acid bacteria on transfer of plasmid between the E. coli strains was determined.

Fermented milk consumption had little effect on the populations of R388 or pUB2380 transconjugants in the anexic mice but the milk fermented with both LAB resulted in a slight increase in E. coli PG1 (R388) transconjugants. Long-term consumption of milk fermented with both LAB strains resulted in inhibition of occurrence and maintenance of E. coli PG1 (pCE325) transconjugants in the anexic mice. L. bulgaricus-fermented milk resulted in a reversible reduction in the numbers of E. coli PG1 (pCE325) transconjugants in mouse faecal samples. When L. bulgaricus and S. thermophilus were grown in BHI broth, the transconjugant lowering capabilities observed with the LAB plus fermented milk were reversed. Administration of L. bulgaricus and the combination of both strains grown in BHI broth increased the population of E. coli PG1 (pCE325) transconjugants recovered from mouse faecal samples. Thus, the authors concluded that probiotic LAB have an effect on the formation and subsequent maintenance of transconjugants in the gastrointestinal tract of anexic mice, and that this effect was dependent on whether the LAB were administered in fermented milk or labouratory growth media, indicating that bacterial metabolic end products as well as viable LAB may be involved.

Clearly much work remains to be done to elucidate the role played by specific environmental parameters, such as diet, ratio of donor and recipient cells, activity of donor and recipient and antagonistic effects of the human gut microflora on plasmid transfer and persistence of GMOs which may be ingested in a viable form in fermented foods. Genetic engineering is now being employed by the food industry to improve the performance of microorganisms used in the production of fermented foods. Concern has been expressed over the exposure of consumers to genetically modified microorganisms which may be ingested in a viable form in future food products. Of chief concern is the risk of recombinant DNA transfer from ingested GMOs to members of the human gut microflora. A range of natural DNA transfer mechanisms are available to microorganisms in the environment. Chief among these are conjugation, transformation and transduction.

Apart from one case of transformation in the oral cavity, most studies examining DNA transfer in the human gastrointestinal microflora have been concerned with conjugation. In vitro and in vivo models of varying degrees of complexity have been employed to simulate the prevailing ecological conditions in the human gastrointestinal microflora. Such studies have demonstrated the transfer of plasmid DNA, both naturally occurring and genetically modified, in the digestive tract.

Some of the factors thought to affect conjugation between bacteria in the gut microflora have also been studied in vitro and in vivo, *e.g.* metabolic burden of plasmid maintenance, recombinant DNA stability, colonisation ability of GMOs and antagonistic activities of the microflora, presence or absence of antibiotics, cell concentrations of bacteria involved in conjugation and the effect of dietary supplements on DNA transfer events in the human gut microflora.

In order to provide information of direct relevance to risk assessment of the use of genetically modified microorganisms in food, further scientific investigations employing in vivo and in vitro models.

Design of GMOs and Colonization of Gut by GMOs

GMOs considered for release into the environment in a viable form will employ several measures to limit the ability of recombinant DNA to be transferred to indigenous bacteria in the target environment. Recombinant DNA may be borne on non-conjugative vectors, on the chromosome of the host strain or by the use of suicidal genetic elements.

However, inbuilt safeguards which prevent the transfer of recombinant DNA under labouratory conditions cannot be relied upon to prevent transfer of recombinant DNA in the natural environment. Thus studies employing in vivo models of the human gastrointestinal tract have been conducted with such GMOs in order to determine their ability to persist in the digestive tract and transfer DNA. Gruzza *et al.* investigated the ability of genetically modified L. lactis strains to colonize gnotobiotic mice.

The genetically modified L. lactis strains were constructed with antibiotic-resistance marker genes inserted on different replicons, chloramphenicol resistance was inserted into a self-transmissible plasmid pIL205, erythromycin resistance was inserted into non-self transmissible plasmids pIL252 (low copy number) and pIL253 (high copy number) and a second erythromycin-resistance determinant was implanted into the L. lactis chromosome.

All strains carried a naturally occurring pIL9 plasmid encoding the machinery for lactose fermentation. It was observed that anexic mice were colonized with the recombinant strains or their plasmid-free derivatives, with plasmid-free derivatives becoming dominant in all cases. Plasmids pIL9 and pIL205 were lost from donor strains but plasmid-bearing parental strains and plasmid-free derivatives were maintained at co-dominant levels, suggesting equilibrium between plasmid loss and conjugation in the mouse gastrointestinal tract.

L. lactis strains bearing the low-copy number non-self transmissible plasmid pIL252 rapidly lost their plasmid complement in the gut of anexic mice, probably due to high segregational instability of this plasmid observed in vitro. L. lactis strains bearing pIL253, the high-copy number, non-self transmissible plasmid colonized the digestive tract of anexic mice but only became dominant when plasmid pIL205 was not present in the same cell, indicating a degree of incompatibility between the two plasmids both plasmids belong to the same incompatibility group. The erythromycin-resistance determinant borne chromosomally was maintained stably in the digestive tract of anexic mice. The authors concluded that in vitro determinations of recombinant DNA stability may highlight if a particular insert is extremely unstable but does not give reliable data on the in vivo stability of the construct.

This, highlights the need for in vivo studies both on the stability of recombinant DNA for efficient GMO function and the effect instability has on the persistence of recombinant DNA in the digestive tract, which is a major factor governing the extent to which DNA transfer between ingested GMOs and members of the human gut microflora occurs.

Duval-Iflah *et al.* investigated the mobilization of pBR-derived recombinant plasmids from donor E. coli K12 strains to E. coli strains of human origin in vitro and in the digestive tract of anexic and gnotobiotic mice associated with the human gut microflora. The genetically modified plasmids used were designed to be non-conjugative (mob-, tra-) and either mobilizable or non-mobilizable. E. coli strains isolated from human faeces were screened for their ability to mobilize the oriT- and oriT+ plasmids.

In anexic mice, associated with donor and recipient E. coli strains, mobilization of an oriT+ pBR derivative by trans-complementation with the gene products of tra (encoded by conjugative plasmid R388) and mob was observed. Such mobilization was only observed sporadically with one E. coli strain of human origin in di-associated mice and not at all in HFA mice. It was determined that about 50per cent of E. coli strains of human origin were able to mobilize oriT- pBR derivatives in in vitro matings but not in anexic mice or HFA animals.

DNA TRANSFER IN FOOD PRODUCTS

The first food products containing genetically modified material are now available on supermarket shelves throughout the world. Much public debate has arisen concerning the safety of such products and indeed the need for genetically modified foodstuffs in the well-stocked larders of the Western World.

However, because genetic engineering offers such technical advantages to the food industry in the mass production of cheap processed food of predictable consistency and quality, there will be increased commercial pressure to broaden the range of genetically modified foodstuffs available in the marketplace. Similarly, as our confidence in the safety of this new technology grows and our ability to apply genetic engineering to products which offer real benefits to the consumer (*e.g.* safer, more wholesome food) and in the development of 'functional' foods which may well play an important role in human health and disease prevention, consumer acceptance and demand for such products may grow.

However, consumer confidence will rely on rigorous assessment of the various potential risks involved and appropriate safety testing. Microorganisms, particularly lactic acid bacteria, have been used in the production of fermented foods for millennia. Recent advances in genetic engineering allow, for the first time, accurate identification of microorganisms traditionally used in food fermentation and the design of novel strains with improved characteristics. Age-old production problems such as instability of industrially important traits and failure of starter cultures due to bacteriophage attack may now be tackled at the molecular level. It has long been proposed that certain lactic acid bacteria or 'probiotics' contribute greatly to intestinal health and well-being. Similarly, genetically modified lactic acid bacteria have also been proposed for use as oral vaccines. Recent advances in molecular microbial ecology allow the scientific basis of such claims to be determined and genetic engineering will enable the design of probiotic bacteria with specific health-promoting properties. No food products containing live genetically modified microorganisms (GMMs) are available in the marketplace at present.

However, commercial pressure on the food industry and the consumer benefits promised by probiotic strains designed with scientifically proven health-promoting capabilities will encourage their use in fermented foods in the near future. Clearly the biosafety of such products must be rigorously investigated before they become commercially available, particularly in view of the public back-lash towards genetically modified plant material used in the food products in Europe.

Of particular concern when considering the release of live genetically modified microorganisms in food is the possibility of recombinant DNA transfer from genetically modified organisms (GMOs) to members of the human gut microflora. Transfer of DNA between bacteria occurs naturally in the environment and offers prokaryotes a unique means of evolution and adaptation in response to changing environmental conditions.

Interest in DNA transfer between bacteria in the mammalian gastrointestinal tract stems from three main areas:

- The possibility of DNA transfers from GMOs which may be ingested in food to members of the human gut microflora.
- The spread of antibiotic resistance amongst bacteria as a result of gene transfer.
- The emergence of novel human pathogens as a result of transfer of virulence factors or antibiotic resistance determinants between bacteria.

The focus of this review is to discuss the ability of bacteria to undergo DNA transfer in the human gastrointestinal tract with particular reference to the risks posed by genetically modified microorganisms, which may be used in the production of fermented foods such as bread, beer, cheese and yoghurt. We will look at the existing procedures available to monitor DNA transfer in the human gut microflora and investigate some of the factors governing frequencies of such transfer events. These studies will provide information relevant to our understanding of the possibility of the transfer of marker genes used in the construction of genetically modified crops to the bacteria present in the mammalian gastrointestinal tract. First let us look at the mechanisms of DNA transfer available to bacteria in natural environments.

TRANSFORMATION PROCESS AND PIECE OF DNA

Transformation is the process by which a naked piece of DNA from the environment binds to the surface of a competent bacterial cell and is taken up by the bacterium. DNA may then be incorporated into the host genome, depending on the recombinational abilities of the host and the 'foreign' DNA. The ability to translocate DNA across the cell boundary is called competence.

Competence is a specific physiological state of a bacterial cell (genetically encoded in some cases, *e.g.* Bacillus subtilis), which occurs transiently and is restricted to certain stages of the growth cycle. Natural competence has been observed in bacteria from a variety of genera, including Haemophilus, Neisseria, Streptococcus, and Bacillus, Acinetobacter and Pseudomonas spp. and Helicobacter pylori.

Recalcitrant species, such as Escherichia coli, may be rendered competent using a variety of chemical, enzymatic and physical procedures, *e.g.* CaCl2 treatment and electroporation. Such physiochemical conditions may sometimes be prevalent in the local environment of recalcitrant bacteria. For example, Ca2+ concentrations in drinking water sometimes approach the levels used in vitro to induce a state of competence in E. coli.

Competence is not the only bacterial-encoded parameter shown to play a role in natural transformation. In some cases, requirements for specific lengths of DNA, DNA states (double or single stranded) and the presence of specific DNA sequences have been observed. Competent Bacillus and Streptococcus spp. may take up any piece of DNA but only homologous DNA will be maintained in the bacterial genome.

Haemophilus spp., on the other hand, requires the presence of specific 11-bp sequences before DNA uptake occurs. These 11-bp recognition sequences occur at a number of locations on the Haemophilus genome. It has also been reported that a diffusible factor may play a role in the induction of competence in Streptococcus spp. Here, induction of competence was found to be dependent on cell density and the pH of the surrounding environment.

Factors Affecting Transformation in Natural Environments

The presence of naked DNA has been demonstrated in a number of natural environments. Naked DNA may arise in a given habitat via a number of routes. Cell lysis as a result of cell death or the activities of bacteriophage will release bacterial DNA and DNA may be released from actively growing bacteria during certain stages of the growth cycle.

The uptake of DNA from the environment will not only depend on a state of competence in the bacterium but also on the persistence of naked DNA in a given environment. In this respect, different microhabitats will vary greatly in their abilities to either protect naked DNA by the presence of favourable salt concentrations or absorption onto solid supports (*e.g.* the surface of soil particles or food particles in the gut) or to degrade DNA, for example, by active nucleases present in the microenvironment.

Despite the longevity of DNA in the soil and the presence of bacteria with known competence for transformation, gene transfer from plant to soil bacteria appears to be an extremely rare event: for example, transfer of an ampicillin resistance gene from a transgenic potato line to the plant pathogenic bacterium Erwinia chrysanthemi was at a calculated frequency of 2 × 10-17 and from transgenic plants to the soil bacterium Acinetobacter calcoacetinus at a frequency lower than 10-13. However, the uptake and integration of transgenic plant DNA via natural transformation has been shown for Acinetobacter sp., strain BD413 by in vitro marker rescue using DNA from various transgenic plants containing the bacterial kanamycin-resistance gene nptII. Naked DNA enters the human gastrointestinal tract from a number of sources. These include ingested food and foreign microorganisms as well as members of the human gut microflora.

Very little is known about the ability of naked DNA to persist in the gut and evade the activities of mammalian and bacterial nucleases. Factors which may affect the persistence of naked DNA, and thus the incidence of transformation in the human gastrointestinal tract, include pH, salt concentrations, cell densities, local nuclease activities and protection afforded by absorption onto surfaces such as food particles or mucosal surfaces.

Such factors would act at the level of the microhabitat and as such would vary greatly even within specific regions of the gut. It has been shown that the CRY1A protein and DNA from genetically modified maize are digested by simulated gastric juices in vitro. However using an in vitro model of the intestinal tract, van der Vossen *et al.* showed that 6per cent of transgenic tomato DNA survived the stomach and small intestine and concluded that the presence of raw mashed tomato helped to preserve the DNA. Free chromosomal DNA of Bacillus subtilis persists for weeks in milk and dairy produce. Such bacteria develop natural competence and can be transformed with free chromosomal or plasmid DNA in such produce. Schubbert *et al.* showed that the wall of the gastrointestinal tract is exposed to a variety of DNA fragments and remains exposed to DNA fragments of dietary origin for hours after ingestion of the food.

The authors found that upon feeding mice 50 μg M13mp18 DNA, approximately 95per cent of ingested DNA was lost during passage through the stomach. However, phage DNA could be detected by PCR and fluorescent in situ hybridization in peripheral leukocytes, spleen and liver cells as well as the contents of mouse small intestine, caecum, large intestine and faeces. Ingested phage DNA was detected for up to 18 hours after ingestion in caecal contents, for up to 8 hours in DNA from the peripheral blood cells and for up to 24 hours but not 48 hours in DNA from spleen and liver. Mercer *et al.* monitored the survival of recombinant plasmid DNA (pVACMC1) in fresh human saliva. The fraction of naked DNA remaining amplifiable in saliva ranged from between 40 and 65per cent after 10 minutes and 6-25per cent after 60 minutes. Amplifiable plasmid DNA was still present after 24 hours incubation in fresh saliva.

The authors also found that plasmid DNA, which had been exposed to degradation by saliva, was capable of transforming naturally competent Streptococcus gordonii DL1 in filtered saliva. Transformation activity decreased rapidly with the extent of plasmid DNA degradation.

Such studies suggest that DNA released from food or bacteria ingested with food may undergo transformation in not only the oral cavity but other regions of the human gastrointestinal tract. Clearly, much more research is needed on the ability of naked DNA to persist in the lower regions of the human gastrointestinal tract and the extent to which transformation contributes to DNA transfer in the human gut microflora.

MECHANISM OF TRANSDUCTION

Transduction is the mechanism by which DNA may be transferred between bacteria by bacteriophage. In essence, the bacteriophage acts as microbial couriers, picking up DNA from one bacterial chromosome and delivering the heterologous DNA to another bacterial chromosome.

Where degradation of naked DNA is one of the chief factors limiting DNA transfer by bacterial transformation in natural environments, bacteriophage protects DNA in natural environments via their proteinaceous capsid. Bacteriophages, on the whole, contribute to gene transfer between bacteria by two main mechanisms, namely specialized and generalized transduction.

Generalized Transduction

Generalized transduction on the other hand, occurs when host DNA is packaged into phage particles instead of the phage genome. This mechanism of transduction has been observed in the lytic phage, P1 and P2 of the enterobacteraceae, for example. Upon transduction to a novel recipient, the transferred DNA may either be incorporated into the host chromosome via recombination or where it possesses the means of self-replication, it may replicate autosomally.

Specialized Transduction

Specialized transduction involves the incorporation of a lysogenic phage into the bacterial chromosome. Lysogeny is favoured by conditions of environmental stress such as the limitation of nutrients and probably aids the survival of the phage and host bacterium. Upon excision from the chromosome, elements of host DNA adjacent to the prophage DNA may become excised along with the phage DNA. The host DNA may then become incorporated into the bacteriophage genome and packaged along with the phage DNA.

Thus, when infection of a recipient bacterium occurs the heterologous DNA may become integrated into the recipient's chromosome during lysogeny.

Factors Limiting DNA Transfer by Transduction

Transduction is greatly dependent on the host range of the transducing bacteriophage, which is generally narrow, since bacteriophage infection is dependent on the phage recognizing specific receptor sites on the bacterial cell surface. Some bacteriophage that are able to mediate transduction between different species of bacteria have been described, *e.g.* P1 and Mu.

The frequency of transduction in nature may be much higher than previously recognized, since the number of bacteriophage particles in many environments appears to be much higher than first thought. Whether the transferred DNA is maintained in the new host is greatly dependent on the ability of the bacteriophage to insert itself and the heterologous DNA into the host genome and evade the bacterial restriction modification system. Bacterial restriction enzymes may recognize specific sequences on heterologous or phage DNA. It has been proposed that restriction modification systems may reduce infection of unmodified phage DNA by 2-3 orders of magnitude. The amount of DNA that may be transferred by transduction is also limited, being about equivalent in size to the phage genome itself.

Transduction in the Human Gastrointestinal tract

Despite the fact that the bacteriophage are ubiquitous members of natural microbial ecosystems, little is known about their distribution or activity in the human gut microflora. To date no reports of transduction in the human gastrointestinal tract have been presented. However, where sufficient effort has been made to isolate bacteriophage from natural environments and examine their activity in situ, they have been found to play a significant role in microbial ecology. The fact that about 90per cent of all bacteriophage isolated from natural environments are temperate suggests that transduction may be more important in microbial genetic plasticity than previously appreciated. Thus specifically designed studies involving in vitro or in vivo models of the human gastrointestinal microflora may well provide evidence of transduction in this ecosystem and elucidate some of the ecological factors governing transduction. Bacteriophage specific for major groups of bacteria present in the gut microflora, *e.g.* Bacteroides spp., bifido-bacteria, lactobacilli, methanogens, clostridia, enterobacteriaceae, streptococci and staphylococci have been identified. However, transduction has not been observed in many of these bacterial groups *e.g.* Bacteroides spp. or Clostridium spp.

Bacteriophage of the lactic acid bacteria play a major role in the dairy industry both as destructive agents, causing the failure of starter cultures in cheese and yoghurt production, and as valuable genetic tools in the development of genetically altered industrial strains. The variety of bacteriophage associated with the lactobacilli suggests that transduction may be a significant means of gene transfer within these genera.

Tohyama *et al.* demonstrated that the Lactobacillus salivarius temperate phage PLS-1 mediates generalized transduction of auxotrophic markers (lysine, proline and serine) and lactose metabolism at frequencies of 10-7 to 10-8 transductants per CFU in vitro. Later, Raya *et al.* showed that phage ADH replicates in a lytic cycle, establishes lysogeny, confers superinfection immunity on the host and mediates plasmid DNA transduction in L. acidophilus ADH. It was also observed that plaque formation on cell lawns of L. acidophilus NCK102 was pH dependent, with the optimal pH for plaque formation being pH 5.5. Transfer of antibiotic resistance determinants have been reported between strains of Desulfovibrio desulfricans via phage Dd1 mediated transduction at frequencies of 10-5 to 10-6 transductants per recipient.

Transduction has also been observed in methanogens, although not strains found in the human gut microflora. Despite demonstrations that transduction does occur under labouratory conditions, little is known about the significance of transduction-mediated gene transfer in the natural and animal-associated environments.

Important questions remain unanswered regarding the prevalence and survival of bacteriophage in different environments, the limitations imposed on transduction by the host specificity of the transducing bacteriophage and the frequencies at which transduction occurs during phage replication.

GENETIC ELEMENTS IN BACTERIAL CONJUGATION

Plasmid DNA, extrachromosomal, self-replicating genetic elements, may mediate DNA transfer between bacteria via conjugation. Conjugation involves cell-to-cell contact between a donor (plasmid-bearing) and recipient (plasmid-free) cell. Not all plasmids are conjugative but many conjugative plasmids carry environmentally important traits such as antibiotic resistance determinants, virulence factors, novel degradative pathways.Different bacteria or groups of bacteria display different mechanisms of conjugation, a comprehensive discussion of which is beyond the scope of this review. However, a short, generalized description of some of the conjugative mechanisms employed by different groups of bacteria may serve to highlight both the mechanistic diversity of bacterial conjugation and that DNA transfer via conjugation-like mechanisms has been observed in bacteria from a wide phylogenetic background.

Conjugation in Gram-positive Bacteria

Conjugation of plasmids in Gram-positive bacteria differs considerably from those in Gram-negative bacteria. Two main mechanisms have been described. Certain plasmids found in Enterococcus spp. and Lactococcus spp. have been shown to initiate conjugation in response to a small peptide signal released extracellularly by plasmid-free recipient cells.

These extracellular peptides act as sex pheromones, and lead to clumping of donor and recipient cell to form a conjugative aggregate. This increases cell-to-cell contact between donor and recipient cells and results in the transfer of pheromone-induced plasmids at high frequencies.

Plasmid pAD1 of Enterococcus faecalis is the best characterized example and is transferred at high frequencies in conjugative aggregates formed in response to pheromones released by plasmid-free recipient cells. The second type of conjugation elucidated in Gram-positive bacteria, especially Streptococcus spp. and Staphylococcus spp., depends on the direct contact between donor and recipient cell imposed by growth on a solid surface.

Plasmids employing this form of DNA transfer are not induced to conjugate by sex pheromones and the mechanism of their transfer remains unclear. Such plasmids often display a remarkable broad host range, *i.e.* they can be transferred to a wide spectrum of both Gram-positive and Gram-negative bacteria, including Streptococcus spp., Staphylococcus spp., Clostridium spp., Lactobacillus spp., Listeria spp., Pediococcus spp. and Bacteroides spp. Many of these plasmids encode one or more antibiotic-resistance determinants and are thought to have played a role in the dissemination of antibiotic-resistance determinants among bacteria of clinical importance. The best characterized examples include pAMβ1 and pIP501.

Conjugation in Archaea

Little is known about the molecular biology of Archaea because of the difficulties in cultivating such organisms routinely under labouratory conditions and the major differences that exist between Archaea and Bacteria. However, a conjugation-like mechanism of DNA transfer has been observed in the halophilic Archaea Halobacterium volcanii. Here, a 'cytoplasmic' bridge forms between parental cells along which chromosomal DNA has been shown to transfer. The mechanism of DNA transfer displays characteristics of both bacterial conjugation and eukaryotic cell fusion. However, until we gain a greater understanding of the molecular biology and indeed microbiology of the Archaea, few conclusions can be drawn about the frequencies or mechanisms of DNA transfer in Archaea.

Conjugative Transposon

Transposons, genetic elements borne by microbial genomes and possessing the ability to 'transpose' between different locations on the same genome, have long been known to play a role in microbial genetic plasticity. Some such elements not only possess the machinery required for transposition but are also capable of mediating their own transfer between the genomes of different bacteria via conjugation. Such conjugative transposons have been found in a wide array of both Gram-positive and Gram-negative bacteria.

They are thought to play a significant role in the dissemination of antibiotic-resistance determinants among a wide range of bacteria and the emergence of multiple antibiotic-resistant pathogenic strains. Conjugative transposons generally excise from the bacterial chromosome to form a covalently closed circular double-stranded DNA transposition intermediate.

The transposition intermediate can either reinsert itself into the bacterial chromosome or onto a plasmid in the same cell, or it can mediate its own conjugation to a second bacterial cell and integrate into the recipient chromosome. Conjugative transposons have a wide host range and have been shown to transfer between bacteria at high frequencies typically 10-5 to 10-4 per recipient.

Conjugation in Gram-negative Bacteria

In Gram-negative bacteria, cell-to-cell contact between a donor and recipient cell is accomplished via the formation of a pilus (which can be short and rigid or long and flexible) by the donor cell. This pilus, a proteinaceous tube, forms a cytoplasmic bridge between donor and recipient, along which a single-stranded copy of the conjugative plasmid is transferred to the recipient cell.

DNA replication then leads to generation of the complementary strand of plasmid DNA in both recipient and donor. Certain plasmids originally found in Gram-negative species display an extremely broad host range. Such plasmids (*e.g.* the R plasmids) may carry a number of antibiotic-resistance determinants and have been shown to transfer to a variety of bacterial species in different natural environments. Some conjugative plasmids may mediate transfer of chromosomal DNA between donor and recipient cells upon integration into the chromosome via homologous recombination and low-fidelity excision from the chromosome, leading to incorporation of segments of bacterial DNA into the circularized plasmid before transfer. Transfer of Gram-negative plasmids has been observed in many natural environments, including soil, rhizosphere, salt and fresh water and effluent.

Mobilization

Some conjugative plasmids may 'mobilize' plasmids co-resident in a bacterial cell. Mobilization results in the transfer of a non-self transmissible plasmid from one bacterial cell to another and is mediated by the conjugative plasmid or conjugative transposon. Mobilization may occur as a result of the non-conjugative plasmid 'hijacking' the conjugative machinery of the conjugative plasmid or the two plasmids may form a co-integrate plasmid. Plasmids with oriT sequences may be mobilized upon formation of efficient cell-to-cell contact between donor and recipient bacteria mediated by the trans acting products of tra and mob genes borne on self-conjugative plasmids co-resident in the same cell. Plasmids devoid of oriT sequences may be mobilized by formation of a co-integrate with a self-conjugative plasmid co-resident in the same cell. Such co-integrate formation is characteristic of bacterial transposable elements. The co-integrate once transferred to a recipient strain via conjugation, may then resolve into its constituent plasmids. Mobilization of genetically modified plasmids has been demonstrated from introduced genetically modified donor strains to bacteria present in different environments. Such observations have implications for the release of genetically modified microorganisms into natural environments, even when steps have been taken to limit the transfer potential of recombinant DNA.

Many broad host range plasmids are also mobilizable, which aids their dissemination within microbial consortia as well as the dissemination of antibiotic resistance often borne on such plasmids. Some conjugative plasmids can also mobilize chromosomal DNA between different bacteria. Conjugative transposons have also been shown to mediate mobilization of non-conjugative plasmids or chromosomal genetic elements between bacteria.

TRANSFER OF PLASMID DNA IN RETROTRANSFER

Transfer of plasmid DNA between bacteria is usually unidirectional, *i.e.* from donor to recipient. However, Mergeay *et al*. demonstrated the transfer of genetic markers from a recipient strain to the donor strain during conjugation experiments. This form of plasmid transfer has been called retrotransfer. The mechanism involved has not been fully elucidated, but is thought to involve multiple donor and recipient cells. Retrotransfer has been Demonstrated in a number of bacterial species.

Factors governing DNA Transfer

Plasmid transfer by conjugation is determined by characteristics of the plasmid itself, *e.g.* tra+, mob+ and oriT+ genotypes, and the donor and recipient bacteria involved. Environmental parameters which affect pilus formation or the rate of plasmid curing, such as SDS, organic solvents, proteases, high temperature and acridine orange, will affect the frequency of conjugation. The metabolic character of the host bacterium, *e.g.* growth rate or stress, may also affect the frequency of conjugation. The frequency of conjugation may be enhanced by the presence of surfaces and bacterial aggregation (*e.g.* mediated by bile salts or sex pheromones). The presence of solid surfaces, *e.g.* soil particles, food particles or walls of mucosa, in an environment may also affect the frequency of plasmid transfer depending on the mode of conjugation employed, by increasing cell-to-cell contact between donor and recipient cells. Substrate availability may also affect conjugation with increased frequencies of conjugation sometimes observed with a plentiful supply of bacterial substrates.

Physiological conditions extant in the environment, such as pH, temperature, substrate availability and bacterial cell density, may all play a role in governing the frequency of conjugation in a plasmid- and bacterium-specific manner. Plasmid maintenance by bacteria exacts a metabolic cost on the cell, *i.e.* the energy needed for the plasmid to replicate and express plasmid encoded proteins.

This 'metabolic burden' may be counteracted by adaptational advantages endowed by plasmid maintenance. Plasmids often encode environmentally important traits such as resistance to antibiotics and bacteriophage, unusual metabolic degradative pathways, virulence factors and genes encoding anti-bacterial compounds. Such traits may enable a plasmid-bearing cell to out-compete a plasmid-free cell in response to altering environmental conditions, *e.g.* presence of antibiotics, unusual substrates or bacteriophage.

Similarly, environmental factors affecting the competitiveness or growth rate of a bacterial cell will affect the ability of the cell to both undergo conjugation and maintain newly acquired plasmids. Properties of the plasmids themselves may limit the extent to which conjugation occurs in the environment. For example plasmid incompatibility, host range, surface exclusion, evasion of host restriction modification systems and segregational stability may all play a role in the extent to which a particular plasmid may spread and persist in natural microbial communities.

Thus it appears that the frequency of plasmid transfer in a given environment will be a function of the interaction between environmental factors and the bacterial populations involved, characteristics of the plasmid itself and the mode of conjugation employed by the plasmid.

FOODS AND DRUG ADMINISTRATION

In May 1992, the US Food and Drug Administration, responding to corporate pressures to remove the prospect of regulation of genetically altered foods, issued a set of rules which have largely left responsibility for protecting the public health and safety in the hands of the industry, permitting products to be marketed without scrutiny unless the industry indicated to the agency that it believed governmental oversight was justifiable.

However, the proposal as published by the FDA in the Federal Register perversely noted several important problem areas: creating allergens, additions of genes from sources which might violate religious and cultural norms (of vegetarians, Jews or Muslims, etc.), and implications for animal welfare and well-being (for example, the incorporation of human growth hormone-producing gene into a pig's genome produced a highly arthritic animal).

Indeed, the commissioner of the FDA, David A. Kessler, and his colleagues noted the possibility that genetically engineered food might 'contain high levels of unexpected, acutely toxic substances'. In order to begin to address ethical issues presented by the genetic modification of foodstuffs, we need to have some ideas of the purposes for which these modifications are performed. (For example, if we wish to approach an analysis on utilitarian cost-benefit grounds.) Despite a great number of general statements, human nutrition and hunger do not appear to be the actual driving forces behind the development of genetically engineered foods.

Other than claiming increased shelf life (which can be indirectly related to nutrition in the sense of reducing spoilage or the consumption of foods that have begun to spoil), there is hardly any indication at all that genetic engineering is being directed to create nutritious substances out of Non-nutritious ones; similarly, there are few real instances of genetic engineering reducing the cost of foodstuffs, or increasing the quantities of foods actually available to populations that are hungry because of the non-existence of consumables (as opposed to hungry because they lack money to buy sufficient foods, no matter how produced). Indeed, early genetic manipulations seemed perversely designed to minimize the achievement of such goals: the creation of 'herbicide tolerant plants' which do not reduce the applications of dangerous chemicals to agricultural fields (which might occur if genetic manipulations of food crops were designed to make them resistant to insects, fungi, or disease) but instead permit higher levels of chemical application, or the introduction of recombinant bovine growth hormone to produce more milk at a time when developed countries suffer from milk gluts and have instituted programmes to kill cows and physically dump milk (milk as a commodity is price inelastic; quantity increases do not lead to price decreases). The goals of genetically engineering foods seem to be more closely tied to increasing the economic gain and power of the corporations involved in food production (for example, the development of herbicide-tolerant farm crops has been led by corporations which manufacture herbicides), making production easier for corporations, reducing or altering packaging and transportation costs for the corporations, etc.

ENVIRONMENT

Industrial societies around the world are faced by massive pollution legacies from the previous emphases on chemical and nuclear industrial activities. The inability of the US Department of Energy to find a suitable site for a longterm nuclear waste repository illustrates our very poor track record for dealing with the scientific, technical, economic, and socio-political aspects of prior technological 'revolutions'.

The environmental problems posed by genetically engineered organisms are likely to be substantially more intractable than those posed by these earlier instances of pollution because genetic wastes multiply, migrate, and mutate. A genetically engineered organism once free in the environment is impossible to recall. (Impacts which are irreversible must be considered with higher scrutiny than those which can be undone.)

Examples of environmental problems which need to be assessed include the risks of transgenic crops themselves becoming weeds; the risk of gene flow to wild relatives which might become weeds or pests; the growth of antibiotic resistance in species (particularly animals), since resistance genes are used as markers for the genetic engineering sites; the problem of exotic or non-native species taking over ecological niches (such as gypsy moth, starling, kudzu, rabbits in Australia, etc.) leading to extraordinary economic losses as well as ecological ones; the restriction (rather than increase) of biodiversity by selective advantageous breeding of transgenic organisms as opposed to natural ones, or the predatory results of expanding exotic species (such as the Dutch Elm micro-organism destroying the American Elm and severely restricting the biodiversity of certain areas in the Northeast United States); health issues (both of plant and animal species, as well as humans - worker health and safety as well as community security); and the occurrence of the completely unexpected, the inability to eliminate uncertainty (for example, the late 1993 floods in the Mississippi valley included the flooding of a field of genetically engineered corn and the dispersal of this plant material to unknown sites within the thousands of square miles of downstream flood plain). Prospective ecological assessment is not being performed, despite the fact that its need has long since been recognized. And economic priorities, particularly those accruing directly to the promoters, as well as the nebulous spur of 'competition' with foreign countries, oftentimes overwhelms any interest in doing environmental impact analysis.

FOOD SAFETY ISSUES

Although the variety registration of existing crop plants has not resulted in adverse effects in humans, the conventional assessments carried out by plant breeders was not considered to be sufficient to ensure food safety of GM crop plants. As a result, Europe issued a regulation on novel foods and novel food ingredients that sets the legal framework for the market introduction of genetically modified organisms (GMOs). This came into force. The Novel Foods Regulation establishes a system of mandatory pre-market approvals of novel foodstuffs, including GM plants.

The accompanying EU guideline gave a clear indication of the types of data that are needed to form the basis of an assessment. In particular, four central safety issues have been set forth related to food and feed safety:

- The nutritional and toxicological consequences of inserted gene products;
- The potential of pleiotropic (unintended) effects in the host organism due to the insertion event;
- The allergenicity of expressed proteins and novel foodstuffs;
- The potential of gene transfer to human and animal gut flora.

This means that in transgenic insect-resistant Bt tomatoes encoding the C-terminal truncated Bt2 gene derived from a Bacillus thuringiensis (Bt) strain, IAb5, the novel protein has to be evaluated on its own merit, while the safety and wholesomeness of the 'remaining' transgenic tomato is assessed separately. Whereas, for example, in antisense RNA exogalactanase tomatoes, which have improved rheologic characteristics due to downregulation of the endogenous exogalactanase activity, the safety assessment is primarily focused on the 'remaining' novel fruit. In general, there is basic agreement on the safety issues to be addressed.

One particular area of concern is the possibility of unexpected or unintended metabolic perturbations due to genetic modification that may alter, for instance, levels of nutrients and health-influencing components. The use of rDNA techniques does not necessarily result in fundamental changes in crop plants compared to the food produced by conventional breeding methods. However, it should be emphasized that a uniform international agreement for the evaluation of GMOs is urgently needed.

For example, an agreement on how to establish in practical terms the similarities or differences between the 'remaining' GM crop plant and the traditionally bred plant seems to be much more difficult to achieve. This chapter is devoted to the description of advanced strategies for the evaluation of potential unintended effects in GM crop plants. Pleiotropic (unintended) alterations in agronomic traits or composition may arise from insertion mutagenesis or as a result of metabolic effects of the novel gene product(s).

The challenge is to gain the ability to discriminate between metabolic alterations due to somaclonal variations, unintended effects or natural diversity. Thereto, emphasis is given to a platform of technologies designed to identify changes in the molecular machinery of crop plants. As is the case for detection methods for GM crops and derived products, there are needs and criteria for sampling strategies, statistical models and manufacturing standard reference matehers recently.

UNINTENDED EFFECTS

Strategies for assessing the food safety and wholesomeness of novel foodstuffs are currently in the exploration phase. Above all there is a need for uniform and harmonized quantitative methods for the testing of potential unintended effects in whole foods, such as GM crop plants are. The safety of, for example, synthetic chemicals and food additives is established by assessing separate elements of the compound in question.

Although very successful in case of single chemicals, the evaluation of a whole food product may not be a simple addition of individual assessments of the constituents. Traditional testing protocols involving labouratory animal 90-day feeding trials with (whole) food products are far from ideal. This type of animal feeding study with complex food matrices is complicated by the likelihood of nutritional imbalances leading to dietary problems, confounding factors, an insensitivity for specific endpoints, and the impossibility of using large safety margins, if any. Therefore, in a global context several regulatory, research and industry-based bodies have proposed alternative concepts for the safety evaluation of genetically engineered foods and food ingredients. In particular the Organisation for Economic Co-operation and Development (OECD), the Food and Agriculture Organization (FAO) and the World Health Organization (WHO) have developed concepts and principles for the safety evaluation of GMOs and derived products.

Their reports concluded that risk assessments should be directed at demonstrating that a GM crop plant or derived food product is as safe as a traditional or previously authorized product. Accordingly the EU Competent Authority incorporated these general principles into their Novel Foods Regulation and accompanying guidelines. In this context, novel foods and food ingredients that are considered to be 'no longer equivalent' to traditionally bred plants are subject to labelling as defined in the Council Regulation.

Substantial Equivalence

The term 'substantial equivalence' appears in the EU Novel Foods Regulation as a major principle in the risk assessment. It emphasizes that, within the context of the Regulation the evaluation of the safety and nutritional value of a GM plant should be focused on a comparative analysis with conventionally bred products. OECD's Group of National Experts on Safety in Biotechnology enunciated this comparative approach towards the safety assessment of transgenic food plants as the concept of substantial equivalence.

In their report, it is assumed that existing food organisms possess a long history of safe use (*i.e.* GRAS). Consequently, the most practical approach to the determination of safety is to consider whether the GM plants are substantially equivalent to analogous food product(s), if such exist. Once substantial equivalence of a novel foodstuff has been established, it provides assurance of safety that is equal to or better than that of its comparators. Further clarification of the concept of substantial equivalence originates from a WHO sponsored workshop. It has been concluded that the establishment of substantial equivalence is not a safety assessment per se, but a dynamic, analytical exercise. Thus, it may mean that analysis of, for instance, gene expression patterns, global changes in protein expression and/or differences in metabolic capabilities (the metabolome) of the novel product in comparison with conventionally bred products should be performed in order to examine equivalence.

This strategy has worked satisfactorily for the assessment of the first-generation crop plants for which sufficient background knowledge is available. However, relatively less experience has been gained with the safety and wholesomeness evaluation of novel food plants for which no substantial equivalence can be established. Even in the case of single gene modification, potential alterations in metabolic pathways are difficult to identify and, consequently, the implications of the genetic modification process on the metabolism of plants are poorly understood. Moreover, data on the mechanisms by which plants regulate, for instance, their response to environmental stress factors and other conditions are remarkably scanty.

Critical Nutrients and key Toxicants

Application of the concept of substantial equivalence appears in practice to lead to various interpretations. Different requirements for risk assessment with regards to the potential of unintended effects exist as a result of genetic modification. Actual approaches entail a consideration of the characteristics of host and donor organism, phenotypic properties and toxicological evaluation and include comparative determinations of any changes in critical (anti-)nutrients and key toxicants for the food source in question.

In general, the analysis of an expanded spectrum of components is unnecessary, but should be considered if there is an indication from other traits that there may be an unintended secondary effect of genetic modification. However, compositional analyses based on single parameters using a list of crop-specific critical (anti-)nutrients and key toxicants as a minimum requirement of screening for possible variations in the content has its limitations.

Limited data are typically available for the less important food components, as in less well-known crops most critical (anti-)nutrients and natural toxicants will be unknown. Moreover, the toxicants to be assessed may be partly influenced by the cellular function of the product encoded by the inserted gene. Furthermore, the natural variation of critical comparators may mask the evidence of secondary effects due to genetic modification.Thus, there is no criterion that determines whether a difference between a GM plant and its natural counterpart is outside the literature values for the ranges of parameters that express natural diversity. Relevant information on indicators of unexpected effects may not have been observed during the development of an edible transgenic crop species. In the future other compounds could become of importance, such as anti-oxidative, oestrogenic and anti-carcinogenic compounds.

In determining critical nutrients and/or key toxicants, differences in consumption patterns and practices of processing and consumption in various geographical settings and cultures must also be recognized. It is our opinion that conclusions about relative safety and wholesomeness based on a list of selected single components may not be equally valid in a global context and in all times. Therefore, there is an urgent need to set up generic approaches using platform technologies to evaluate undesirable metabolic perturbations in GM food crops.

A Post-genome Challenge

In our view the identification of unintended effects encompasses the ability to interpret and use innovations in molecular genetics research such as genomics, proteomics and metabolomics on a crop-by-crop basis. This will establish a whole data package directed towards a holistic view on possible unintended side effects due to genetic modification. In essence, the focus is shifting towards molecular characterization to understand functional activity.

Vital to this approach will be informative profiles of molecules at different integration levels, *e.g.* mRNAs, proteins including post-translational modifications and molecules at the level of primary and secondary plant products and metabolites that might be useful benchmarks for the detection of unintended effects. The relative importance of these various 'profiling' techniques in establishing data sets for assessing unintended effects is not indicated by their order and will vary from species to species. Alterations in, for example, expression levels, post-transitional modification, interactions or chemical composition do not per se imply that the product is less safe or even unsafe. Such a platform of technologies may determine whether unintended effects exist between the GM plant and its control(s) or whether there is substantial equivalence apart from certain well-defined expected differences.

Measures of the relative activities of various molecular constituents associated with metabolic capabilities in non- and modified plant cells under different conditions will provide new insights into how metabolism and genetic modification are orchestrated. But it should be emphasized that the fields of plant genomics and proteomics are still in their infancy.

8

Genetic Control of Interspecific Differences

THE GENETIC IDENTITIES OF SPECIES

The degree of divergence in genes controlling reproductive and ecological systems is not necessarily representative of the total genetic changes that have ensued during speciation. Therefore, one would not expect a consistent relationship between taxonomic categories and levels of genetic identity based on loci that were not involved in reproductive isolation.

In some instances the genetic identities of closely related species may be similar to those of conspecific populations.

Consider the case of *Gaura longiflora* (Onagraceae) and its presumed derivative *G. demareei*. Gottlieb and Pilz found the same mean genetic identity of 0.99 for pairwise comparisons of populations of the same and different species.

Similarly, the mean genetic identities for populations of the closely allied Heuchera villosa (Saxifragaceae) and H. *parviflora* are 0.98 and between H. americanum and *H. pubescens* 0.99, the same as conspecific populations. Gottlieb et al. reported mean population identities for the Layia discoidea (Asteraceae) and its progenitor, L. glandulosa, of 0.90 compared to 0.93 for conspecific populations.

The aforementioned values are based on allozyme loci. High levels of identity have been found between *Mimulus guttatus* and its derivative, M. micranthus, for both allozyme and chloroplast DNA restriction fragment length profiles. High levels of identity using both approaches also have been found between *Clarkia biloba* and its derivative, C. lingulata.

High genetic identity between species on continents often is not associated with substantial divergence in morphology or ecological profile. However, high identities on oceanic islands often are associated with considerable morphological and ecological divergence. In the Hawaiian composite *Tetramolopium*, the genetic identity among species is 0.95. In Hawaiian *Bidens*, the mean identity is about 0.95. These values contrast with a mean genetic identity of ca. 0.65 for typical species on continents. The breeding system, through its effect on the genetically effective sizes of species, has an important bearing on the genetic identities of species. In the Hawaiian *Schiedea* (Caryophyllaceae) and its close ally *Alsinidendron*, predominantly self-fertilizing species have lower genetic identities with other species in either genus than do predominantly outcrossing species with similar levels of morphological divergence. Outcrossing allogamous species with small population sizes also tend to have relatively low genetic identities.

Allozyme data often are assumed to be valid descriptors of the overall level of variation within species and genetic divergence between species. This assumption seems valid, given there is typically a good correspondence between allozyme data and other forms of variation, including morphological. Incongruities between allozyme and other data in Neo-species ostensibly signify their rapid evolution through selection and genetic drift.

Given that most species traits tend to remain relatively static over time, we would expect allozyme divergence to catch up with the traits by which the species are diagnosed. When genera are considered in their totality, there typically is a good correspondence between allozyme and morphological divergence in the sense that species within different sections of a genus are more divergent in both respects than species within the same section.

It is evident that major trait differences may be controlled by one to a few major genes or several QTLs, each with minor effects, or a combination of major and minor genes. Of particular interest is the possibility that there may be common genes involved in the same character differences between different pairs of species in the same genus.

Our understanding of the pleiotropic effects of plant genes is woefully inadequate. Theory predicts that genes are apt to have negative pleiotropic effects, and that these effects will retard or oppose a response to selection. The data base is too small even to know whether negative pleiotropic effects are common. Speciation involves a change in several traits. These transitions may be accomplished by changes in a small number of loci. The number that actually is involved in a given instant remains to be determined. Given that there are a number of character differences between species, we may ask whether the transitions occur synchronously or consecutively. If the differences involve a functional character suite, the transition may be synchronous or nearly so. If the differences involve independent functions, then a temporal relationship is not required.

The Ecogenetic Species Concept calls for ecological divergence and genomic disharmony. As genomic disharmony typically is a by-product of ecological divergence, except when chromosomal changes are involved, it is likely that ecological changes promote the development of genomic disharmony, rather than the other way around. In this respect speciation is both a product and a by-product of adaptive evolution. When chromosome rearrangements are involved as well, then speciation also becomes a product of stochastic evolution, because the fixation of novel underdominant rearrangements typically is based on chance. In this case, genomic disharmony via translocation or inversion heterozygosity may precede an ecological shift in the course of speciation.

We may expect little relationship between the magnitude of character differences between species and the level of differentiation displayed by allozymes or other ostensibly neutral markers. This is supported by the observation that rapid radiation on oceanic islands may yield very divergent species with high levels of genetic identity, whereas moderately divergent species of ancient vintage may have lower levels of genetic identity.

CROSS-INCOMPATIBILITY

Interspecific cross-incompatibility is a complex process involving the timing and level of pollen grain germination on heterospecific stigmas, and the rate and length of tube growth and its ability to penetrate the micropyle.

Interspecific cross-incompatibility may be controlled in part by the same locus (S) that controls self-incompatibility. Alterations in S-alleles and in S-allele expression by modifier genes have been correlated with shifts in interspecific cross-compatibility in *Nicotiana* and *Lycopersicon*. In *Lycopenicon* the major control of unilateral incompatibility between *L. esculentum* and *L. hirsutum* (crosses are successful only when the former was the egg parent) can be attributed to the S locus, but two additional loci on other chromosomes also affect crossability.

Simple genetic control by other loci also has been reported. Two genes (Kr_1 and Kr_2) have been implicated in the partial crossing barrier between *Triticum* and *Secale*. A single dominant gene conditions partial incompatibility between *Hordeum ttulgare* and *H. bulbosum*. In contrast crossincompatibility among conspecific populations of *Phacelia dubia* (Hydrophyllaceae) is polygenic. Which of these genetic architectures is the norm remains to be determined.

HYBRID STERILITY

The sterility of hybrids may be due to the interaction of cytoplasmic and nuclear factors in gametes, the recombination of genes belonging to disparate genomes during meiosis, abnormal pairing of chromosomes during meiosis, or the failure of reproductive organs to form in a normal manner. Given that this discussion is concerned with the genetic bases for interspecific character differences and genome incompatibility, the nuclear-cytoplasmic basis for hybrid sterility will be considered first.

Discord between nuclear and cytoplasmic genes is manifested in reciprocal crosses. In *Oryza sativa* (rice) hybrids between subspecies *japonica* and subspecies *indica* spikelet sterility is 17per cent when the former is the female parent versus 51per cent when the latter is the female parent. In hybrids with ssp. *japonica* as the female parent, pollen sterility is 38per cent versus 57per cent when *indica* is the female parent. In some instances, reciprocal-specific effects are associated only with male function. Hybrids between *Phacelia dubia* var. *dubia* and P. dubia var. 'railroad' have an average of 90per cent pollen fertility when the former is the egg parent versus 65per cent fertility when the latter is the egg parent.

When male function varies as a reciprocal-specific effect, crosses in one direction often yield hybrids that are male sterile, whereas crosses in the other direction yield hybrids with moderate to normal male function. Typically one or two nuclear genes are involved in male sterility. Male sterility in hybrids between *Nicotiana langsdorffi* and *N. sanderae* is due to two genes, each of which occurs in two allelic forms. In hybrids between *Helianthus annuus* and *H. grosseserratus*, male sterility is conditioned by one gene whose expression is environment-dependent.

Recently interest has turned to the location in the cytoplasm of the elements causing male sterility. In *Epilobium* there is altered mitochondrial gene expression in male sterile hybrids, which suggests that mitochondria are involved with male sterility. Chloroplast gene expression is unaltered in these hybrids. There are insufficient data to say whether mitochondrial gene action is associated with male sterility in hybrids of other genera.

In most plant hybrids, sterility is associated with differences in chromosomal homology or genie disharmony in the haploid generation. Some hybrids have normal meiosis, but are sterile nevertheless. Many hybrids have abnormal chromosome pairing.

This can be due to the action of specific genes that affect the stringency of chromosome pairing, differences between species in the amount, nature and dispersion of repetitive DNA and interspecific differences in chromosomal arrangements. Stebbins contends that the latter is the decisive factor in determining the sterility of F_1 hybrids.

The chromosomal basis for reduced hybrid fertility has been described in hybrids between several pairs of progenitors and Neo-species. *Clarkia lingulata* differs from its putative progenitor *C. biloba* by a translocation and a paracentric inversion. The former has a haploid number of 9 versus 8 in the latter. Hybrids have less than 15per cent viable pollen.

Chaenactis glabriuscula (Asteraceae) ostensibly has given rise to *C. stevioides* and *C. fremontii*. The parental species has a haploid chromosome number of 6 versus 5 in both derivatives. The derivatives differ from the parental species by at least two translocations. These translocations are not shared by the derivatives as seen by abnormal chromosome pairing in their hybrids. Indeed *C. stevioides* and *C. fremontii* differ by two translocations. The fertility of hybrids between *C. glabriuscula* and each of its derivatives varies from about 10 to 50per cent. Chromosomal differences need not be large to cause a reduction in fertility. Stebbins notes that many sterile hybrids with normal or nearly normal chromosome pairing were rendered fertile by artificial doubling of chromosome number. Had sterility been under genic control, it would have persisted in the doubled hybrid. In general, fertility declines as the number of chromosomal differences between species increases. Large rearrangements have a greater impact on fertility than very small ones. Chandler, Jan, and Beard sought the relationship between the number of chromosomal differences and pollen fertility in 34 types of interspecific hybrids (F_1 s) in *Helianthus*. Hybrids heterozygous for three or more translocations have lower fertility on average than those with none or few. The correlation between the number of translocations and pollen fertility is $r = 0.30$, which falls short of statistical significance.

Pollen fertility in *Helianthus* also is diminished by inversion heterozygosity and the occasional failure of chromosomes to pair. The mapping of quantitative trait loci in intersubspecific rice hybrids has uncovered supergenes that may be "cryptic structural rearrangements," specifically inversions, on two chromosomes. Supergenes are groups of tightly linked genes within which recombination will cause reduced fitness.

Pollen sterility is associated with heterozygotes and ostensibly occurs as a result of recombination within the supergenes. The mapping of 117 loci in hybrids between *Tetramolopium humile* and *T. rockii* (Asteraceae) reveals a high degree of segregation distortion, which may be due to cryptic structural hybridity. The hybrids show no reduction in fertility. The nature and magnitude of chromosomal rearrangements that contribute to hybrid sterility and by which species are differentiated have been elusive because of a paucity of chromosomal markers. The utilization of a broad range of genetic markers now permits the rather dense genetic mapping of plant chromosomes.

From such maps one can discern chromosomal rearrangements of various sizes. Most comparative chromosome mapping has involved species of agronomic importance (e.g., *Lycopersicon* and *Capsicum*; *Sorghum* and *Zea*; *Triticum* and *Hordeum*). Such mapping reveals that species in different genera may differ by several translocations and paracentric inversions. Even divergent species within a genus may differ by several rearrangements. One of the first applications of chromosome mapping in the study of hybrid sterility involved *Lens culinaris* and *L. ervoides*. The species were thought to differ by a single translocation. Using a segregating F_2 population, Tadmor, Zamir, and Ladizinsky were able to correlate four isozyme markers with quadrivalent formation during meiosis and identify the translocation end points. Plants with pollen viability below 65per cent are heterozygous for the translocation, while plants with viability above 85per cent are homozygous, thus implicating the translocation as the primary basis for reduced hybrid fertility.

A study of the same type was conducted in *Helianthus*. *Helianthus annuus* and its close ally *H. argophyllus* differ by two reciprocal translocations. Quillet et al. analysed the segregation of isozyme, RAPD and morphological markers in backcross (BC_1) progeny. Over 80per cent of the variation in pollen fertility is explained by genetic intervals located on three linkage groups. Meiotic abnormalities are tightly correlated with the markers circumscribing these intervals, indicating that the translocations were the prime cause of reduced hybrid fertility.

Of special interest here is the chromosomal architecture of species and their derivatives. The most informative data set is on *Helianthus*, and involves *H. annuus*, *H. petiolaris*, and their hybrid derivative, *H. anomalus*. Rieseberg et al. generated linkage maps based on 212 loci in *H. annuus* and 400 loci in *H. petiolaris*. The genetic markers were mapped to 17 linkage groups, corresponding to the haploid chromosome number of the three species. By comparing the genomic location and linear order of homologous markers, Rieseberg et al. were able to infer chromosomal structural relationships among the three species.

Helianthus annuus and *H. petiolaris* differ by at least 10 separate structural rearrangements, which include three inversions and at least seven reciprocal translocations. First-generation hybrids between these species are semisterile. Pollen viability is less than 10per cent and seed viability is less than 1per cent. *Helianthus anomalus* did not simply combine the chromosomes of its parental species. Rather it has undergone a minimum of seven novel rearrangements, including a minimum of three chromosomal breakages, three fusions, and one duplication. First-generation hybrids with *H. annuus* have pollen fertilities of 2 to 4per cent and those with *H. petiolaris* have pollen fertilities of 2 to 58per cent.

AMOUNT OF GENETIC CHANGE

What can we conclude about the amount of genetic change necessary for the evolution of a new species? This depends on our species concept. If reproductive isolation, or divergence in fertilization systems, alone are the criteria for speciation as in the Biological Species and Recognition Concepts, respectively, then relatively few genes may be sufficient for species formation. If the criterion for speciation is divergence in both reproductive and ecological systems as in the Ecogenetic Species Concept, then genetic change of a greater magnitude is required. The number of genes responsible for species differences at the time of their divergence almost certainly is much higher than the minimum number sufficient for speciation, because ecological divergence and/ or reproductive isolation is not likely to have been achieved by the smallest number of gene changes. Moreover, speciation may have been precipitated by a number of character or genomic changes acting in concert.

FLORAL CHARACTERS

Major genes and minor genes can both be responsible for interspecific differences in floral attributes. The classic study on the floral morphology of *Aquilegia* by Prazmo illustrates the effects of major genes. Based on F_2 segregation patterns, he found that one locus determines whether spurs are straight or curved, a trait that differs among species groups.

A second locus controls whether the flower is erect or nodding, the latter being dominant. Erect flowers are visited by lepidopterans, whereas nodding flowers are visited by bees and hummingbirds. A pair of duplicate loci governs whether the flower has a spur on each of its five petals. Spurs are the dominant condition.

Species of the Asteraceae sometimes differ by the presence/absence of ray florets as in *Layia discoidea* and *L. glandulosa*. The former is discoid and a likely derivative of the radiate *L. glandulosa*. Ray presence/absence is governed by two independently assorting loci. In *Haplopappus* the presence/absence of ray florets is under single-gene control.

This difference may impose a partial ethological barrier between the species, because pollinators are attracted more to radiate variants than to discoid ones in *Senecio* and *Bidens*.

Bradshaw, Wilbert, and Schemske demonstrated that major genes are involved in the floral divergence and isolation of the bee-pollinated *Mimulus lewisii* (*Scrophulariaceae*) and the hummingbird-pollinated *M. cardinalis*.

Flowers of *M. cardinalis* have red petals lacking nectar guides, a tubular corolla, a large volume of nectar, and exerted anthers and stigmas. Flowers of *M. lewisii* have pink petals with yellow nectar guides, a wide corolla orifice, a small volume of highly concentrated nectar, and inserted anthers and stigmas. Using DNA polymorphisms and genome mapping techniques, Bradshaw et al. demonstrated that each difference in petal anthocyanins and carotenoids, corolla width, petal width, nectar volume, nectar concentration, and stamen and pistil length is determined by at least one gene of major effect—that is, a gene accounting for more than 25per cent of the phenotypic variance among F_2 plants. This study and conventional analyses of F_2 segregation patterns show that the petal colour difference is controlled by one gene and that other differences are controlled by several genes. Studies on *Mimulus* also demonstrate that some floral traits are under complex genetic control. *Mimulus cupriphilus*, a predominantly selfing species, ostensibly is a recent derivative of the mixed mating *M. guttatus*.

The former has smaller flowers and no petal spots. Based on estimates of minimum gene number from F2 segregation patterns, at least three to seven genes control differences in flower size, and at least three genes control flower spot number. Differences in corolla lobe shape are controlled by one major gene.

Another recent derivative of *M. guttatus*, *M. micranthus*, also is predominantly selfing. The evolution of small flower size in the latter is associated with more rapid floral growth and shorter duration of floral development. Segregation patterns in the F_2 generation indicate that the species differ in 8 and 10 factors for the growth rate and duration of bud development, respectively. The selfer is characterized by smaller corolla width and reduced pollen production. The minimum number of loci differentiating the species for these traits are 8.7 and 13.5, respectively.

Geum coccineum and G. rivale produce hybrids that are fully fertile and exhibit normal chromosome pairing. In *G. coccineum* the sepals are reflexed and the petals are horizontal. In *G. rivale* the sepals and petals are erect. The F_2 segregation pattern was skewed toward the former, but nevertheless suggested polygenic control for floral orientation.

The petals of *G. rivale* are 6-10 mm wide, while those of *G. coccineum* are 12-27 mm wide. The binomial regularity of the F_2 progeny for petal width is indicative of polygenic control. The difference between the cream-coloured petals of *G. rivale* and the orange-red petals of *G. coccineum* is due to two pairs of complementary genes and perhaps additional modifiers.

Finally, the genetic basis for interspecific differences in some quantitative floral traits has been studied in *Nicotiana*. Differences in corolla size and shape, and differences in the ratio of corolla limb to tube length in the bud and their relative rate of elongation are polygenic.

Ecological Characters

The genetic architecture of ecological differences between related species is poorly understood. Presumably some combination of major and minor genes are involved. Major genes have been implicated in heavy metal tolerance in several species. The difference in habitat preference of *Mimulus guttatus* and its copper-tolerant derivative *M. cupriphilus* ostensibly is controlled by a major gene and modifiers.

We may look at the basis for alternate edaphic preferences within species for clues to explain differences between species. Consider *Silene vulgaris*. Two distinct major genes are responsible for zinc tolerance, and two major additive genes also appear to control copper tolerance in this species.

The genetic basis of alternative flowering time has been studied within *Arabidopsis thaliana*. Kuittinen, Silanpaa, and Savolainen mapped quantitative trait loci (QTLS) for flowering time in the progeny of a cross between an early-summer and an overwintering Finnish population. They found that the trait is controlled by one major and six minor QTLS, the former explaining 53per cent of the variance in flowering time.

Change from the perennial to annual habitat has accompanied speciation in many genera, but the genetic basis for the change is poorly understood. Some insights are available in *Mimulus*. Macnair and Cumbes crossed the perennial *M. guttatus* and derivative annual *M. cupriphilus*. The F_1 hybrids are perennial, thus indicating dominance for this expression. The annual habit segregates in advanced generation and backcross hybrids. The number of genes involved could not be determined. Many close relatives have divergent growth forms because of differences in apical dominance. The domestication of crop plants often has involved an increase in apical dominance, and here we have some insights into its genetic control. The profound increase in the apical dominance of maize relative to its probable wild ancestor, teosinte, is due largely to a single gene (*teosinte branched 1*). The maize allele of *tbl* is expressed at twice the level of the teosinte allele.

There is considerable interest in the genetic control of rhizomatousness, as this trait has a bearing on the weediness of species. One weed that spreads very effectively by rhizomes is johnsongrass, *Sorghum halepense*, an allotetraploid. The control of rhizome production was determined in the progeny of crosses between its putative diploid parents, *S. bicolour* and *S. propinquum*, by Paterson, Schertz et al. They found that this trait is influenced by eight QTLS; gene action in four is additive, in two is dominant, in one is recessive, and in one is overdominant.

The mapping of quantitative and other traits onto chromosomes reveals surprising similarity in the genetic architecture of shared traits among members of the same taxonomic families. A close correspondence among QTLs affecting seed size, disarticulation of the inflorescence, and day-neutral flowering time has been shown for rice, barley, wheat, sugarcane, sorghum, and maize. A correspondence among a high proportion of genes affecting height has been shown for maize and sorghum. As more information is forthcoming, we will see whether QTL analysis in one taxon typically has much predictive value for other taxa within different genera of the same family. Extensive conservation of gene repertoire and order seems to be the case in cultivated cereals, and in several genera including *Brassica*, *Helianthus*, and *Solanum*.

Given the conservation of gene repertoire and to a large extent gene order, it is possible that parallel changes associated with speciation in different parts of a genus are dictated by changes at the same loci. It is possible that the multiple and independent shifts from a perennial to an annual habit in *Gilia* are controlled by the same loci. Support for parallel changes at corresponding genetic loci comes from studies on the genetic control of large seeds, Non-shattering seeds, and day-length insensitive flowering in sorghum, rice, and maize.

These similar phenotypes are largely determined by a small number of QTLs that correspond closely in location in the three species. Correspondence in location does not prove that the genes are identical, but it suggests that some of them may be. This idea is bolstered by the tendency of corresponding QTLs to have similar gene action.

THE TEMPO OF SPECIATION

The time necessary for species divergence is not a simple function of the number of genetic or chromosomal changes involved in the process. Although speciation involving relatively few changes may be rapid, it is not necessarily the case. The population size and strength of selection are also important.

Selective change in several traits may be sequential instead of simultaneous. If some potentially adaptive genes have negative pleiotropic effects, time may be required for alterations at modifier loci. Stochastically driven change may be very slow, especially if multiple chromosomal changes occur.

The time necessary for the development of postzygotic barriers due to a pair of complementary genes was treated by Orr and Orr in relation to species subdivision. If alternate alleles were fixed at two loci as a result of genetic drift, the time was not affected by the number of isolated populations that constitute a species.

However, if the complementary genes were pleiotropic and affected by natural selection, the reproductive isolation of populations occurs the quickest when a species is split initially into two large populations. Conversely, when postzygotic barriers require stochastically driven change such as chromosomal rearrangements, then reproductive isolation occurs the fastest when population size is small.

Insights into the speed of postpollination barrier emergence may be gained from studying barriers between selected lines, conspecific cultivars, and cultivars and their wild ancestors. Cultivars pass through genetic bottlenecks and experience periods of inbreeding and directional selection, as do populations undergoing major ecological transitions. Consider the case of *Phlox drummondii*, which is an endemic annual of central Texas. Seeds were collected by Thomas Drummond in 1835 and subsequently distributed to nurserymen in Europe. Over the next 80 years, over 200 true-breeding lines were established. They differ in corolla size, outline and colour, and plant branching pattern and stature. The "Grandiflora" strains were derived from the original material and were the source of the "Dwarf" strains. "Extra Dwarf" strains appeared on the market somewhat later. The most recent strains are 'Twinkles', which are derived from the "Extra Dwarf" strains.

Levin measured the level of cross-incompatibility between wild Phlox and cultivars. Cross-incompatibility arose as a by-product of cultivar evolution; the greater the number of "phylogenetic steps" between phloxes, the greater is the level of cross-incompatibility.

There is a 14per cent reduction in crosscompatibility when wild *Phlox* is crossed with "Grandiflora" strains, a 20per cent reduction when it is crossed with "Dwarf" strains, a 35per cent reduction when it is crossed with "Extra Dwarf" strains, and a 50per cent reduction when it is crossed with 'Twinkle' strains, relative to cross-compatibility within wild *P. drummondii*. Compatibility was measured in terms of the percentage of pollen grains sending tubes into the pistil.

Whereas rates of barrier building may be obtained for cultivars with known histories and pedigrees, very little information is available for natural populations. One intriguing situation is in *Mimulus guttatus*. Lindsay and Vickery studied the cross-compatibilities and hybrid fertilities of populations in the Bonneville Basin of Utah using as a natural time clock habitats provided by the recession of glaciers and lakes.

The present habitats differ in temperature and length of the growing season. Populations in habitats less than four thousand years old differ among each other in morphology and display substantial reductions in seed-set and reduced hybrid fertility when crossed with other populations. Thus rapid diversification was accompanied by rapid barrier building.

The key question that remains is the tempo of speciation—that is, the time required for both ecological change and the development of at least partial postpollination reproductive barriers.

The answer is quite straightforward in the case of allopolyploids, where a shift in ecological profile and barrier building are by-products of hybridization and chromosome doubling. The time is a few generations. This has been demonstrated in the course of hybridization between diploid *Tragopogon* species and the formation of persistent tetraploid populations.

The stabilization of diploid hybrid derivatives also may be relatively rapid. Simulation analysis by McCarthy et al. indicates that recombination speciation is punctuated, with a rapid transition to a stabilized entity following a period of stasis in a hybrid swarm. That period may be hundreds to thousands of generations, depending on the proximity of parental species, breeding system, and the relative fitness of hybrids.

Fertile, stable plant Neo-species have been synthesized experimentally in several plant genera, even when F1 hybrid fertility is low. One experiment involved *Nicotiana langsdorffi* and *N. sanderae*. The F1 hybrids are semifertile, with pollen fertility near 50per cent. Three selection lines were derived from the moderately fertile F2 population, one favouring short corollas like *N. langsdorffi*, another favouring long corollas like N. sanderae, and another intermediate corolla tubes.

By the 10th generation, the three lines had been stabilized for the selected flower size, several unselected floral and vegetative traits, and normal fertility. The small-flowered line has normal fertility, whereas the other two are partially sterile. The lines have fertility barriers between them. Hybrids between the small-flowered derivative and N. sanderae have pollen fertilities near 60per cent, and hybrids between this derivative and *N. langsdorffi* have fertilities between 46 and 58per cent.

In another experiment Grant crossed *Gilia malior* and *G. modocensis*, and obtained a hybrid that was almost completely sterile. Its pollen fertility averaged 2per cent and its seed fertility averaged 0.007per cent. Ten generations of inbreeding, and selection for vigour and fertility, yielded a line that was fully fertile with a unique combination of parental traits, and that was intersterile with the parental species. It is noteworthy that the vast preponderance of F2 to F6 plants were sterile, weak, or both. The chromosome number of the parental species and F1 hybrid is $2n = 36$. Chromosome numbers in the F2 generation vary from $2n = 37$ to 40. The stabilized derivative has $2n = 38$. Two other lines selected for vigour and fertility reverted to one parental type.

Ungerer et al. tested the hypothesis of rapid hybrid speciation in *Helianthus* by estimating the number of generations of recombination to stabilize the *H. anomalus* genome, which was derived from *H. annuus* and *H. petiolaris*. They found that stabilization could occur within 60 generations. The rapid transition may be achieved in spite of the fact that F1 hybrids are mostly sterile, because in synthetic hybrid lineages pollen fertility recovered to 90per cent in only four generations of sib mating or backcrossing. This study demonstrates that recombinational speciation in *Helianthus* could have been rapid, not that it indeed was rapid.

Speciation involving chromosomal rearrangements and ecological shifts typically is spoken of as saltational or quantum. The speed of the transition remains to be determined. The critical issue here is the time necessary to fix novel chromosomal rearrangements. Chromosomal evolution is dependent on stochastic processes, which typically will not propel a population with the speed associated with selection. Then one requires the confluence of chromosomal change and ecological change.

If populations undergoing chromosomal change are small as inferred, the amount of variation in them may be insufficient to support a robust response to selection. Stebbins suggested that ecological and chromosomal change can be completed during a few hundred or thousands of generations.

Speciation associated with an ecological shift and little chromosomal evolution could occur even faster than so-called quantum speciation. Diversification unopposed by serious competitors, such as sometimes occurs on oceanic islands or edaphic islands, may render new species in a few hundred to a few thousand generations.

The extensive and rapid radiation of some genera on oceanic islands may not be as surprising as it seems. Major habitat shifts (*e.g.*, the invasion of metalliferous soils) have occurred within the past few hundred years in many plant lineages, as have changes in breeding systems and flowering phenologies.

NATURE OF GENETIC DIFFERENCES

Consider next the nature of the genetic differences between ancestor and Neo-species. Charlesworth modeled the fixation of mutant genes in partially self-fertilizing Neo-species, and concluded that they would tend to differ from their ancestors in traits that were recessive, because partial selfing facilitates the fixation of advantageous recessive mutants. Dominance relationships of diagnostic traits inferred from crosses between ancestor and partially selfing Neo-species support this expectation. If selfing facilitates the fixation of recessive mutants, then we might expect that recessive traits would play a proportionally smaller role in the divergence of outbreeding Neo-species.

Another reason recessive alleles will have a smaller role in the divergence of outcrossers is that recessive alleles are less likely to spread in response to natural selection than dominant alleles. In inbreeders there is no theoretical barrier to the spread of recessive genes.

The genetic architecture of character changes has a bearing on the rate of character change. Characters governed by a small number of genes and having a high additive genetic variance tend to spread faster than characters governed by many polygenes and having low heritability and major epistatic effects.

Orr and Coyne also note that a mutation's probability of fixation is directly proportional to its phenotypic effect, and that a major gene contributes more to the character under selection than does a minor gene. This is important because the short-term response to selection depends more on the relative magnitude of gene effects than on the specific number of loci.

PLEIOTROPY AND LINKAGE

Mayr implicitly refers to pleiotropy in the establishment of reproductive isolation in allopatric populations. He states that "the ecological shifts in incipient species are bound to have an effect on their isolating mechanisms. The thesis of the origin of reproductive isolation as a by-product of the total genetic reconstitution of the speciating population is consistent with all known facts."

Given that changes at loci may have manifold effects, there are limitations on their incorporation within populations. Genes of major effect often have large negative pleiotropic consequences and are not likely to be incorporated within a population without corresponding changes at modifier loci. Thus we should not assume that one allele of major effect will simply (and quickly) replace another.

In view of the prevalence of pleiotropy, we would expect many examples in plants. However, this is not the case. Our best understanding of pleiotropy for the same trait across species involves genes governing anthocyanin synthesis. Plants homozygous for an allele conferring white petals in *Phlox drummondii*, *Echium plantagineum*, and in

Digitalis purpurea have reduced vigour and competitive ability. The pleiotropic effect of a mutation at an anthocyanin locus is indicated in chimeric plants, where white-flowered sectors show less growth than pigmented sectors. The converse also may be true. In Ipomoea purpurea white flower homozygotes at the W locus are vegetatively larger and produce more flowers than darkly pigmented homozygotes. The pleiotropic effect of "white" alleles at anthocyanin loci may be due to the accumulation and diversion of precursors to alternate pathways. One of the best examples of pleiotropy involves the S locus in *Nicotiana* tabacum. An allelic substitution that affects petal size and outline also affects leaf, calyx, anther, and capsule dimensions. Another striking example of pleiotropy is in Arabidopsis thaliana, where flowering-time genes affect leaf length, number of rosette and cauline leaves, and number of axillary flowering shoots of the main inflorescence.

In *Lycopersicon esculentum* the gene for suppressed lateral branches also suppresses apical growth and corolla development, and causes deformed anthers and reduction in flower number per inflorescence. Alteration of the genetic background through hybridization modifies the pleiotropic effect of the "suppressed lateral" gene. The intensification of the *L. pimpinellifolium* genome through backcrossing brings about a rapid advance Towards the normality of the corolla.

A shift Towards normality in one trait is not necessarily accompanied by such a shift in other traits. This study is important, because the effect of the genetic background on pleiotropy is almost unknown in plants.

Pleiotropy or linkage cause segregation distortion for flower colour in *Mimulus*. *Mimulus cardinalis* has red flowers and M. lewisii has pink flowers. The difference is controlled by a single gene with pink being dominant to red. When grown in an environment where few plants die, F_2 progeny of the interspecific cross approximate a 3 : 1 ratio of pink- to red-flowered plants. However, when the F_2 were grown at the Stanford garden in the summer, a 1.7 : 1 pink: red ratio was obtained. Conversely, when a subset of the same progeny were grown at timberline, an 8.7: 1 ratio was obtained. Nobs and Hiesey also grew a replicate of the F_2 s in the Stanford garden in the winter and observed a 4.4 : 1 ratio of pink to red.

The distorted segregation ratios were associated with high seedling mortality in environments that are hostile to one or the other species. *Mimulus cardinalis* occurs at low elevations in the California mountains; *M. lewisii* occurs at high elevations under much cooler conditions.

High seedling mortality occurs in *M. cardinalis* seedlings when grown at high elevations; such mortality occurs in *M. lewisii* at low elevations. Mortality in F_2 progeny and distorted phenotypic ratios in warm environments ostensibly involves the selective elimination of *M. lewisii* types in warm environments and of *M. cardinalis* types in cool environments.

Evidence for the linkage of genes controlling the contrasting floral traits of related species is forthcoming from *Lycopersicon*. QTLs controlling reproductive behaviour and floral traits were mapped in a backcross population of *Lycopersicon esculentum xL. hirsutum*. The locale on chromosome 1 that contains the S (self-incompatibility) locus also harbors most major QTLs for flower size, flower number, bud type, and inflorescence rachis length. The presence of the H allele at the S locus is associated with more flowers per plant, larger flowers, and longer inflorescences. The S locus also controls interspecific cross-incompatibility. Stigma exsertion is unlinked to the S locus. DeVicente and Tanksley also mapped a QTL for flower number per plant to the S locus in progeny from a *L. esculentum* x *L. pennellii* hybrid. The association between S locus and floral traits occurs in other families and may be indicative of a common, conserved ancestral gene complex.

Another interesting example of linkage has been demonstrated in the recent divergence of the weedy, late-developing *Senecio vulgaris* ssp. vulgaris from its Non-weedy, early-developing parent, *S. vulgaris* ssp. *vulgaris*. Genetic studies reveal that the major gene for developmental speed is linked to the ray-floret gene, as are chromosomal regions that influence leaf and numbers, and pedicel length.

This chromosomal region exhibits a substantial proportion of Non-additive gene effects, which contrasts with the traditional additivepolygenic model of evolutionary divergence. A similar pattern occurs within subspecies of *Plantago major*, where the *Pgm*-1 locus is associated with many loci affecting ecologically important traits. We may now ask how often adaptation and speciation involve major genes. Orr and Coyne contend that major genes are involved to a greater extent in adaptation and as a basis for interspecific differences than we might judge from theory or qualitative arguments. They reviewed theoretical analyses of macromutational and multifactorial evolution and found that they tend to be biased in favour of the polygenic control of novel adaptations as espoused by Fisher.

The usual argument against the importance of major genes in adaptation and speciation is that they are likely to have harmful pleiotropic effects that prevent fixation in natural populations, even though these effects may be overcome by intense selection. Orr and Coyne conclude that there is little empirical support and inconclusive theoretical support for the notion that adaptation and speciation almost always involve genes of small effect. The data presented here support Orr and Coyne's position.

Genes affecting quantitative trait variation have been mapped in populations derived through interspecific hybridization. Such mapping provides the much-needed insights into the genetic basis for species differences that cannot be obtained from classical breeding studies that rely on means and variances from a cross between two populations. Not only will gene numbers be estimated, but it will be possible to discriminate among many genes of small effect and major genes with their modifiers.

HYBRID WEAKNESS AND LETHALITY

The lethality or weakness of some hybrids is due to the complementary action of alleles at one or a few loci. Perhaps the best-known example of an aberrant developmental pattern in hybrids caused by a single gene is the "corky" syndrome expressed in some hybrids between *Gossypium hirsutum* and *G. barbademe*. Young stems and occasionally leaf petioles and mid-ribs develop a corky surface.

The syndrome also includes the suppression of apical dominance, profuse branching, and dwarfism. The "corky" syndrome is based on the genotype ck^x ck^y; the former allele comes from G. *hirsutum* and the latter comes from G. *barbadense*.

First-generation hybrids between *Oryza breviligulatus* and *O. glabenima* often are weak. This condition is controlled by two complementary dominant genes, W_1 and W_2 that are common in *O. breviligulatus* and *O. glaberrima*, respectively. Two complementary dominant lethals are responsible for chlorosis in hybrids between varieties of *O. sativa*. Complementary genes causing hybrid inviability also occur in *Crepis* and *Gossypium*.

In many hybrid combinations, the F_1 plants are vigourous and fertile, but segregates in the F_2 are either weak or partially sterile. Advanced-generation breakdown usually has been attributed to genic imbalance across numerous loci. However, breakdown may have a simple genetic basis.

In *Setaria italica* where some 172 hybrids have poor growth, the segregation of normal to weak was 15: 1. This indicates that weakness may be due to two complementary recessive genes. Chlorosis in F_2 hybrids of crosses between some *Melilotus* species is based on two sets of duplicate recessive genes.

Two complementary recessive genes also are responsible for a reduction in vigour in F_2 hybrids derived from different strains of *Oryza sativa*. Of particular interest is their approximate location in the genome. Fukuoka, Namai, and Okuno mapped each gene independently in F_2 populations derived from two strains. One gene (*hwd 1*) was mapped onto the distal region of chromosome 10, and the other gene (*hwd 2*) was mapped onto the central region of chromosome..

Crosses between many combinations of species yield various levels of seed abortion. Seed abortion is due to disharmonious interaction among the three types of tissue in the developing seed—that is, maternal tissues, embryo, and endosperm. The genetic basis for hybrid seed abortion is poorly understood. One informative pair of studies was conducted on *Lens*. Ladizinsky, Cohen, and Muehlbauer obtained an interspecific hybrid between *L. ervoides* and *L. culinaris* using embryo rescue techniques.

This plant produces half normal and half aborted embryos upon selfing. Pod abortion ranges from 10 to 90per cent in the F_2 progeny, implying quantitative control. A later study indicated that the abortion of F_1, hybrids is strongly affected by dominant or epistatic gene interaction. There is no evidence that the aforementioned genes involved in hybrid weakness or inviability have any adaptive significance. Thus it is of interest to consider the basis for inviability in some hybrids between a copper-tolerant species, *Mimulus cupriphilus*, and its Non-tolerant progenitor, *M. guttatus*.

Two independent genetic systems afford partial reproductive isolation. One gene that confers copper tolerance, or another tightly linked to it, interacts with a small number of genes in *M. guttatus* to give seedling death in hybrids in the fourth-leaf stage or later. Another gene that confers copper tolerance, or another tightly linked to it, produces hybrid seedling death before the first true leaves expand.

CONCLUSION

The genetic changes leading to the formation of new species are poorly understood and are the subject of considerable debate. Some evolutionists propose that major gene changes may propel populations into different reproductive or ecological realms and thereby promote speciation. Speciation may be a direct result of selection acting on these genes or a pleiotropic consequence of adaptive shifts elsewhere in the genome. Others propose that the interactive effects of many minor gene loci produce divergent reproductive or ecological profiles.

These views are not mutually exclusive; some differences may be dictated by major genes, while others may be dictated by polygenes. I will not evaluate the merits of each of the views. Indeed all of the aforementioned genetic changes are involved to various degrees in speciation. Rather, the genetics of reproductive and ecological discontinuities will be considered on a character by character basis, recognizing that species will differ by more than one character. The genetics of genomic incompatibility, including cross-incompatibility, and hybrid weakness and sterility also will be considered.

The genetic basis of differences between species that confer pre- and postzygotic barriers to gene exchange has been difficult to determine because crosses often are incompatible or hybrids are inviable or sterile. Because the strength of pre- and postzygotic barriers are positively correlated with the degree of genetic divergence, the genetic basis for differences differentiating species will be best understood for closely related species. If we analyse species that are distantly related, we may overestimate the number of genes initially involved in reproductive isolation, because most of the change may have followed the speciation event.

Indeed most of the following examples do not involve recent progenitor-derivative pairs, so their value in understanding speciation genetics is somewhat reduced. Nevertheless, the forthcoming estimates are of value because the number of genes controlling a trait, the magnitude of their effects, and their position in the genome affect the speed and mode of divergence.

9

Plant Genetic Diversity, Sustainable Agriculture and Variation

THE EXCHANGE OF INFORMATION AND GERMPLASM

The value of all plant genetic resources lies in their utilization in crop improvement programmes or in other bioindustries. The vast majority of conserved germplasm is available for distribution to the global scientific community but this international distribution poses considerable risks of accidental introduction of non-indigenous plant pathogens and pests.

The detection and elimination of such disease-causing agents is therefore essential.

Often, such agents will be present at very low levels and may actually be symptomless. In this context then, new technologies reviewed in Chapter have been of great value in the development of reliable and sensitive nucleic acid- or antibody-based diagnostic probes, and these will be of great importance in optimizing the health and value of germplasm collections.

Finally, if the biotechnology revolution has had great impact in developing novel molecular and cellular technologies, no less a revolution has affected radically the way in which the world organizes and disseminates information.

Powerful computers linked by the Internet have facilitated the development of sophisticated and comprehensive informatics systems which enable information on plant genetics and molecular biology to be cross-related to systematic, ecological and other data through international networks. Chapter explains how this information is organized and distributed against a background of continual development.

NOVEL TECHNOLOGIES FOR BIODIVERSITY ASSESSMENT AND MONITORING

The analysis and characterization of genetic diversity is fundamental to any *ex situ* or *in situ* conservation strategy. Over the years the methods for detecting genetic diversity have expanded from Mendelian analyses of discrete morphological variants, to biometrical approaches, biochemical techniques based upon protein and isozyme profiles, to methods based on DNA sequence variation. There is now a huge array of molecular marker techniques, and Chapter reviews the major technologies and their appropriate use in addressing key issues in germplasm conservation The application of molecular technologies to increase our understanding of biodiversity at several levels is exemplified by Chapter, which is concerned with the population level, which is concerned with molecular approaches to genetic relatedness.

Chapter reviews how the collection of diverse germplasm for *ex situ* purposes and the development of effective *in situ* collections both require an understanding of the spatial and temporal aspects of population structure and associated genetic diversity and the factors that control it.

Studies with low- or single-copy markers are particularly valuable in answering questions about genetic relatedness between species. Chapter illustrates how comparative molecular mapping provides new opportunities for exploring the genetical basis of phenotypic variation within and between plant species.

The implication of comparative mapping is that genes controlling specific phenotypic traits will have corresponding map locations in different taxa. This provides a new framework for the identification of homologous genes in germplasm and new tools for the effective utilization of germplasm in crop plant improvement.

This approach is also valuable in the study of quantitative traits controlled by several genes since flanking markers for major genes provide markers for quantitative trait loci (QTL) manipulation and selection and the map-based cloning of at least some genes controlling quantitative traits.

Biotechnological Tools for Conservation

Biotechnology can make a substantial contribution to the quality of germplasm collections and their efficient management. For example, in the management of very large germplasm collections there is a clear requirement for procedures for fast, reliable taxonomic identification of material and the elimination of redundancy. Samples characterized by molecular criteria may prove superior to traditional analyses of phenotype and examples of the use of DNA markers to guide the management of germplasm are explored in Chapter by reference to one of the world's largest collections of germplasm at the International Rice Research Institute.

Not all plant species can be stored in the form of desiccated seed. Socalled 'recalcitrant' species have to be kept in moist, warm conditions and even then have relatively low longevity. In addition, there are many cultivated plants which must be maintained clonally, such as banana, which produce few seeds, or potato, which does not come true from seed. To these traditional problem species we must now add the products of biotechnology, new categories of germplasm including clones of elite genotypes, and transformed cell lines, for which storage of seed is inappropriate. Biotechnology has impact on storage technologies in two quite different ways. The first of these, concerns the various types of *in vitro* culture which allow the propagation of plant material in an aseptic environment, thus ensuring the production of disease-free stocks and simplifying quarantine procedures for the international exchange of germplasm. The development of sensitive molecular probes to detect low levels of pathogens.

A more radical approach to safe storage and distribution of germplasm, and possibly the only way to significantly reduce the rate of loss in biodiverse tropical forests, within the bounds of available resources, is to develop alternatives to the costly (and inherently vulnerable) methods of *ex situ* storage of seed or tissue cultures, by conserving germplasm in the form of DNA or DNA-rich materials.

The latter include those dried specimens conserved in the vast resources of the world's herbaria.

The DNA Bank-Net, an international network of DNA repositories, has now been established with both storage and distribution functions. Although technologies for rapid removal, analysis, amplification and secure storage of small amounts of DNA are now well advanced, research is still needed on methods to amplify the entire genomic DNA of a species, which means that storage of DNA or DNA-rich materials should still be considered to be something of an insurance policy rather than a replacement for conventional modes of germplasm storage.

Biotechnology-assisted Crop Improvement

Effective use of genetic resources in plant breeding programmes requires methods to determine whether useful genetic variation exists, and costeffective methods to introduce useful genes into the material. Biotechnology has made a contribution to both these aspects. The use of genetic maps based on molecular markers linked to genes of interest facilitates the more rapid selection of both simple and complex traits thus accelerating their incorporation into breeding materials. Where traits are controlled by one or a few genes, direct isolation of genes and their incorporation into elite cultivars by increasingly efficient transgenic technologies provide an additional approach to the more traditional methods of plant breeding. This prospect also challenges one of the classical concepts in plant genetic resources work, namely the 'gene pool' concept of Harlan and de Wet (1971), which separated cultivated species and their wild relatives into different categories based upon their 'crossability'. The value of this concept in guiding the work of the plant breeder must now be re-evaluated in the light of molecular genetic technologies which have not only cast a completely new light on the relatedness of crop plant genomes, but which have effectively widened the potential gene pool of a species to include all other life forms. What challenges does this imply for the traditional plant genetic resources programmes? Why is the emphasis on conserving species that are related to crops when more exotic and potentially valuable genes may be present in liverworts, or green algae, or other less studied taxa?

BIOPROSPECTING FOR NOVEL GENES

The value of genetic resources as a source of novel compounds for various industries has been alluded to above. For centuries biodiversity has provided sources of fuels, medicines, foodstuffs, pesticides, etc. The high returns available from introducing a new drug, for example, plus the realization that biodiversity is rapidly being lost (estimated rates of extinction range from 30 to 300 species per day) has encouraged new attitudes to the exploration and exploitation of biodiversity through 'bioprospecting'. How does biotechnology impact on this traditional area? A fundamental operation in any bioprospecting programme is the screening of thousands of extracts for compounds of interest. Gene technology now allows the speeding up of screening programmes for new compounds through the development of more sophisticated *in vitro* assays. For example, the genes encoding receptor proteins for certain classes of drug, or enzymes that may serve as targets for novel pesticides, may be cloned and expressed on a large scale in high-throughput *in vitro* assays into which thousands of plant extracts may be applied. Bioprospecting is also concerned with discovering novel genes and natural variants of genes or gene products that may either be used directly or to guide sequence improvement by molecular methods. Chapter gives examples of amino acid sequence improvement in peptide hormones as a result of screening homologous genes from the wild, and this area of biotechnology is likely to develop in parallel with new molecular methods of generating random variation through employing the mutational potential of the polymerase chain reaction (PCR) – i.e. 'forced molecular evolution' – or recombinatorial libraries of peptides.

Whilst gene prospecting has provided a new perspective on biodiversity, the area is not without political controversy because these advances tend to favour agriculture and other industries of developed countries, and may actually displace traditional commodity production in resource-rich but underdeveloped countries. Chapter also outlines the development of strategies which integrate conservation and exploitation with economic returns and sustainable development in those countries from which the bioresources were first obtained.

PRESENT LIMITATIONS ON DESCRIBING PLANT GENETIC VARIATION WITH MOLECULAR TOOLS

Human genetic research often focuses on genomic mapping and analysing specific genes. In contrast, many plant genetic resource questions are best addressed by evaluating numerous individuals from many taxa.

Thus widespread use of molecular markers in plant genetic research will involve preparing and analysing large amounts of DNA template and assaying many individuals at the same time. However, techniques have not yet been developed to extract, clone, replicate or amplify DNA in nanolitre volumes.

Also needed are techniques for pooling DNA samples from individuals in a population, assaying the pooled sample, and generating a comprehensive picture of population variation. In the future, sequence-based approaches may be useful for these tasks. At present, high costs make these approaches impractical for many high-throughput applications.

PLANT GENETIC VARIATION

Effective conservation and use of plant genetic resources involve asking many questions about the extent and distribution of genetic variation. Only when the appropriate markers and technologies for describing this variation are accessible can such questions be adequately addressed.

The most appropriate markers for a given question should: (i) be heritable; (ii) discriminate between the individuals, populations, or taxa being examined; (iii) be easy to measure and evaluate; and (iv) provide results that can be compared with results of similar studies. Molecular markers and assays that meet these criteria were introduced in the 1950s and have proliferated ever since. Use of protein isozymes as genetic markers increased rapidly after their introduction. Nucleic acid markers gained popularity in the 1970s, with the advent of DNA sequencing and restriction fragment analysis. The number of different molecular markers and assays has increased exponentially since the late 1980s, when introduction of the polymerase chain reaction (PCR) made DNA amplification and sequencing possible and affordable.

As the diversity of markers and assays increases, the cost of using them continues to decrease. At the same time, the range of taxa and tissues that can be analysed has expanded. Improved DNA extraction, amplification, and sequencing protocols allow us to analyse ancient DNA, from 13,000-year-old specimens of *Mylodon* (ground sloths) and 17–20 million-year-old *Taxodium* (bald cypress) specimens, and samples as small as single cells or fractions of a seed.

Thus the chances are good that at least one marker assay will be appropriate for any question about variation in any taxa. Furthermore, results of one study can generally be related to results from many other studies.

Such opportunities are inevitably accompanied by challenges. None of the challenges are unique to molecular studies, but all become more critical as the number of available marker assays increases. Faced with a bewildering variety of molecular technologies (and their acronyms), choosing a marker and assay that are appropriate for a scientist's particular question, taxa of interest, and practical constraints can be overwhelming. The temptation is great to use techniques that are recommended by coworkers, are frequently used, for which the most information is available, or are the newest – without adequately evaluating the techniques to determine which is most suitable.

Because molecular markers can be measured in virtually all plant taxa and minimal tissue is required, extra care must be taken to verify that the entries in a test array are correctly identified. Since molecular markers are widely applicable and their constraints are not always understood, thoughtful planning is particularly important when designing molecular studies, to ensure that they have an appropriate range of entries, an appropriate number of entries per taxon, and conclusions that stay within the scope of a specific question and test array. In addition, the vast amount of published research using molecular markers complicates the task of thoughtfully relating one study to others with similar marker assays, research questions, and/or taxa.

Another challenge is formulating research objectives that ask critical questions about plant genetic resources. Descriptive studies that use molecular marker assays to examine variation between and within specific taxa can generate valuable information, but some plant conservation questions are best addressed by experimental studies. The wide variety of molecular marker assays available provides new ways to address such questions – if the markers are appropriately used. Such challenges can be met successfully and thoughtful decisions can be made if logical frameworks are available to categorize and compare these valuable molecular tools. A solid foundation for such comparisons has been provided recently. This chapter builds on that foundation by reviewing both early and recently developed molecular markers and assays, citing examples of their use. The types of questions asked about plant genetic resources are described, as are guidelines for selecting appropriate markers and assays to address these questions. The literature cited here is far from comprehensive, but includes applications of each marker assay in plant genetic research. While the list of molecular techniques will soon be dated, hopefully the framework presented will be helpful for categorizing and comparing new tools as they are made available. The chapter concludes with a perspective on new marker assays as they apply to, and may change, the study of plant genetic variation.

TECHNOLOGY TRANSFER FROM THE HUMAN GENOME PROJECT

Technology transfer from the Human Genome Project (HGP) to the study of plant genetic resources has been very important in two areas: (i) sequence characterization and marker development; and (ii) integrating all stages in genetic analysis projects, from template preparation through data interpretation. Some of these innovations are now being used in plant genetic research; others will require some level of optimization before they are applied.

Plant genetic variation is best characterized with a large number of highly informative, easily analysed molecular markers. While the number of PCR-based markers will most likely continue to increase, the question at hand is how these markers will be detected. Manual, autoradiographic detection of amplified markers presently predominates in plant genetic research. However, ongoing programmes in human genotyping via semiautomated, fluorescent detection of SSRs are progressing at an incredible pace. Reed *et al.* (1994) reported that semi-automated technology can genotype over 100 000 samples, using approximately 250 marker loci, in less than six months. Moreover, Schwengel *et al.* (1994) found that data generated by this method were at least as accurate, more efficient, less labour intensive, and as cost effective as data generated by standard radiolabelling techniques.

Human genotyping kits are available for approximately 400 markers that define a ±10 cM resolution genomic index map. Kits for plant species and families are in development. Although cost remains an issue, these marker assays are clearly capable of handling the large arrays commonly encountered in plant conservation and breeding programmes.

The costs of generating and detecting numerous marker loci for each of many individuals can be reduced by evaluating several of an individual's loci in a single lane on a gel. In the studies cited above, SSR loci from an individual were multiplexed at the electrophoresis level: several loci were amplified in separate reactions, the PCR products were pooled, and the pooled sample was electrophoresed. In contrast, several laboratories are developing methods to multiplex at the reaction level, amplifying several SSR loci in one tube. Eleven putative loci in canola (*Brassica napus* L.) have been amplified simultaneously, using 11 pairs of fluorescently labelled primers in one reaction. The loci were distinguished from each other by fragment size and fluorescent label colour. This approach is also under way with SSRs of maize and sorghum [*Sorghum bicolor* (L.) Moench]

Another molecular marker used in human genetic research reveals discrete changes (single nucleotide substitutions) in a specific DNA sequence. Variation of these polymorphic sequence tagged sites (pSTS) is the most frequent and widely distributed type of sequence variation in the genome. These individual nucleotide markers are usually biallelic within a population, and are generally less polymorphic than SSRs and other tandem repeats. However, a number of closely linked pSTSs can be combined into a multilocus marker that is as informative as an SSR.

Developing and assaying pSTS markers involve several steps. Random genomic DNA clones (each 300–600 bp long) from selected plant species are sequenced, then single nucleotide differences between individuals are identified by comparing automated sequencing electrophoregrams. Biallelic pSTSs are selected, and primer pairs are designed to amplify them. After a pSTS locus is amplified, the oligonucleotide ligation assay (OLA) is used to label the PCR products and discriminate the two allelic forms. In this assay, the PCR products are denatured and then serve as templates for a ligation detection reaction (LDR). This reaction uses three 'probes': two diagnostic probes, each of which complements one of the two pSTS alleles, and one labelled probe that is common to both alleles.

The diagnostic probes anneal to complementary template sequences, then are ligated to the labelled probe. The two diagnostic probes differ in length. Thus when the LDR-labelled fragments are electrophoresed, pSTS allelic variation between entries can be detected as variation in migration rate of fragments on the gel, due to differences in fragment length. As in many PCR-based marker assays, significant costs are presently incurred for pSTS marker development. Once markers are developed, however, the costs for each assay are not high.

The previously highlighted markers are usually associated with Non-coding regions of the genome. Complementary efforts are under way to develop high-throughput, automated methods to partially sequence cDNA clones, then use them to identify expressed genes in many organisms, including plants.

Thanks to the rapid proliferation of expressed sequence tag (EST) databases, at present the probable gene function of a cDNA clone can often be inferred solely from the similarity of nucleotide or amino acid sequences between the clone and genes of known function.

Using these techniques to compare isoforms of a gene (particularly its untranslated 3' regions) between members of a plant species or family, PCR-based markers that describe variation of this gene could be developed. When assays using these single-gene markers are compared with multilocus studies of the taxon, a more complete picture of genetic variation may be revealed.

TYPES OF QUESTIONS, MARKERS, AND MARKER ASSAYS

All studies of genetic variation have four primary components: (i) the research question; (ii) the marker used to address it; (iii) the technique used to generate and measure the marker; and (iv) the method used to analyse the data produced. This chapter focuses on the first three components.

Types of Plant Genetic Resource Questions

Most questions asked about *ex situ* and *in situ* plant genetic variation fit into one of three categories. While boundaries between them are not always clear, these categories are useful when deciding how to address specific questions. Questions of identity are forensic in nature and ask whether two or more samples are genetically the same. These questions concern whether genebank accessions are present in duplicate, whether populations are genetically distinct or are geographical subdivisions of one population, and

whether genetic change has occurred in an accession or population over time.In contrast, questions of location and diagnostics concern the presence and location of a particular allele or nucleotide sequence in a taxon, genebank accession, *in situ* population, individual, chromosome, or cloned DNA segment. These questions are asked to locate populations or individuals that have desirable traits; determine whether traits have been lost from a population, or have been transferred from one population or plant to another; and establish the position of markers on physical or genetic maps. The differences between diagnostic and forensic questions are significant, but are not always clarified. Both types of question require qualitative information about specific loci. Yet while diagnostic studies often use single-locus markers, each with few alternative states, forensic questions are commonly addressed with information from many multiallelic loci.

Finally, relationship and structure questions examine relatedness or similarity between genotypes, the amount of genetic variation present, and how variation is distributed between individuals, populations, and taxa. Such questions are used to pinpoint gaps in genebank collections, decide which genotypes are high priority for *in situ* conservation, design germplasm sampling strategies and regeneration programmes, construct core collections, study gene flow, and determine optimum sizes for conserved populations or genebank accessions. These issues are generally addressed with information from many loci. Quantitative data can be used, but qualitative data are more appropriate for some phylogenetic and parentage questions.

Types of Molecular Markers

The appropriate markers for a study can discriminate between entries in an array, but are not so polymorphic that important variation is masked by random noise. Molecular markers range from highly conserved to hypervariable, and can be either proteins or nucleic acids. The nucleic acids used as markers include entire genomes, single chromosomes, fragments of DNA or RNA, and single nucleotides. A wide variety of nucleic acid fragments are utilized as markers. While some occur once in a genome, others are repeated. Many repeated sequences used as markers are Non-coding; others are elements of multigene families. Some repeated sequences are interspersed throughout the genome, either distributed randomly or in clusters. These interspersed repeats are common in plant and animal nuclear genomes, and are found in plant (but not animal) mitochondrial genomes. The chloroplast genome contains a large inverted repeat (IR); most angiosperm chloroplasts have two copies, separated by a short single- copy region. Repeat length and (rarely) loss of one copy can vary between taxa.

Much research at present is focused on repeated sequences that occur in tandem. The classes of tandem repeats are distinguished by the length of the core repeat unit, the number of repeat units per locus, and the abundance and distribution of loci. The names for these classes are themselves varied and have been inconsistently used, but Harding *et al.* (1992) and Tautz (1993) have clarified the nomenclature. Tandem repeats were first reported in the literature as 'satellites' of DNA, detected in CsCl density gradients as fractions with different GC content than the rest of the genome.

These satellites have repeat units that are usually several hundred nucleotides long, with thousands of copies at each of several loci in the nuclear genome. These loci are usually in heterochromatin, often near centromeres. Satellite DNA is present in numerous species.

For many satellites, the number of loci and number of repeat units per locus vary between species and higher taxa.Minisatellites (often called variable number of tandem repeat loci, or VNTR loci) are widely used as markers, especially in forensics.

The repeat units are usually less than 100 nucleotides long, with tens to hundreds of copies per locus.

Thousands of loci in a genome may have similar core repeat units. The number of repeat units at a minisatellite locus can vary greatly between individuals and populations. First described in humans, minisatellites are found in numerous animal species, often near telomeres.

They are also common in plants, and are often associated with satellites and centromeres. The number of repeat units per locus is less variable in plants than in animals, but is still high; plant minisatellites are useful markers for variation between and within species.

As suggested by their name, microsatellites – also called simple sequence repeats (SSRs), or simple sequence length polymorphisms (SSLPs) – have very short repeat units, no more than six nucleotides long. SSRs are more abundant than minisatellites in Non-coding regions of the nuclear genome, and are present in some nuclear genes and organelle genomes.

The number of repeat units per locus is lower for SSRs than for minisatellites, but can approach 100 in animals and 50 in plants. The abundance and polymorphism of SSRs make them particularly valuable for describing variation between populations and individuals.Like minisatellites, SSRs were documented first in humans and later in plants. Plants and animals differ in the abundance of specific SSR motifs in the genome.

In both plant and animal genomes, chromosomal distribution of SSRs is variable. Some animal SSRs are found near heterochromatin or interspersed repeats, but most are randomly dispersed. Dinucleotide SSRs are randomly distributed in *Arabidopsis thaliana* (L.) Heynh., but other studies have located plant SSRs near genes, highly methylated

DNA, satellites, or centromeres. Tandemly repeated genes are also utilized as markers.

Perhaps the most widely used are the nuclear genes that encode ribosomal RNA (rRNA). The three rRNA genes are separated by two internal transcribed spacer regions, generally referred to as ITS1 and ITS2. These genes and spacers form a unit that is tandemly repeated hundreds of times, at one to several loci in the genome. At each of these loci, the individual repeat units are separated by Non-transcribed intergenic spacer (IGS) regions. In the middle region of each IGS are tandem copies of a short subrepeat sequence.

Variation in rRNA gene clusters can be measured at several levels, each evolving at a different rate: (i) the number and location of rRNA loci, which is highly conserved; (ii) the (more variable) number of tandem gene clusters per locus; (iii) the conserved sequences of the three genes; (iv) the variable sequences of the ITS regions; and (v) the highly variable number of subrepeats in the IGS region. These features make rRNA gene clusters versatile and informative markers for mapping and phylogenetic analyses.

Types of Molecular Marker Assays

Molecular marker assays are generally classified by whether the molecules evaluated are proteins or nucleic acids, and whether the character analysed in a nucleic acid marker assay is the entire genome, a chromosome, a fragment, or a nucleotide. Alternatively, marker assays can be categorized by the type of character measured. Some methods measure quantitative differences between entries in an array. Others measure qualitative characters, each with two or more possible states. Marker assays also differ in the number of loci evaluated per analysis, whether multiple loci are evaluated simultaneously or sequentially, and the type and amount of information needed about the marker loci before conducting the assay.

Protein Marker Assays

When proteins are used as genetic markers, the assumption is made that any variation between proteins reflects heritable variation in their amino acid sequences.

This assumption has long been debated, since protein phenotypes can be affected by factors such as post-translational modification, plant phenology, and environmental conditions during plant growth. For many tasks, however, protein marker assays continue to be useful molecular tools.

Table. Summary of Molecular Market Assays used to Measure Plant Genetic Variation

Marker assay	Type of molecule	Genomes assayed[a]	Character analysed	No. of character states[b]	No. of loci per assay[c]	Multilocus analysis[d]	Inheritance[e]
Microcomplement fixation	Protein	Total	Reactivity	Quant.	One to many	Sim.	-
Monoclonal antibody assay	Protein	Total	±Reactivity	2	One or several	Sim.	-
Protein electrophoresis	Protein	n, cp	Electromorph	<10	One to several	Seq.	Codom.
CsCl density gradient	DNA	Total	Buoyant density	Quant.	Many	Sim.	-
DNA–DNA hybridization	DNA	Total	ΔT_m	Quant.	Many	Sim.	-
Flow cytometry	DNA	n	DNA content, no. of chromosomes	Quant.	Many	Sim.	-
Chromosome banding	DNA	n	±Band	2	Many	Seq.	Codom.
Fragments electrophoresed, then detected:							
Single-copy or low-copy RFLP assay	DNA, RNA	n, cp, mt	±Restriction site	2	One to several	Seq.	Codom.
Multilocus restriction fragment assay	DNA	n, cp, mt	±Fragment	2	Many	Sim.	Dom.
Arbitrary PCR	DNA, RNA	n, cp, mt	±Fragment	2	Many	Sim.	Dom.
Designed-primer PCR	DNA, RNA	n, cp, mt	Fragment length	≥2	One to many	Seq.	Codom.
Fragments detected directly, without electrophoresis:							
Dot or slot blot hybridization	DNA, RNA	n, cp, mt	±Fragment	2	Many	Sim.	Dom.
			Signal intensity	Quant.	Many	Sim.	-
In situ hybridization	DNA	n	±Fragment	2	One to many	Seq.	Dom.
Nucleic acid sequencing	DNA, RNA	n, cp, mt	Nucleotide	4	One	-	Dom. or codom.

[a]n = nuclear, cp = chloroplast, mt = mitochondrial.
[b]Quant. = quantitative.
[c]In the nuclear genome; the chloroplast and mitochondrial genomes are each considered as one locus.
[d]Loci analysed simultaneously (sim.) or sequentially (seq.).
[e]For loci in diploid (nuclear) genomes, dominant (dom.) or codominant (codom.) inheritance of alleles.

Some assays measure immunological properties of proteins. When antigenic proteins of one plant or animal species are injected into an animal (usually a rabbit), antibodies are produced. Genetic differences between entries in an array can be assessed by comparing how each entry's proteins bind to antibody serum (antiserum) from a reference species. This binding affinity is measured by the amount of antibody–antigen precipitate produced or by microcomplement fixation (MCF). Each MCF solution contains antiserum from a reference sample, antigen from a test entry, and a protein complement. The complement can only bind to antibody that is bound to the antigen. The amount of unfixed complement is measured; this is used to calculate the amount of unbound antigen, which indicates the extent of variation (the immunological distance) between the entry and the reference sample. After MCF assays are conducted for each entry in an array, these immunological distances are compared between entries. MCF has long been used in vertebrate phylogenetic analysis.

Until recently, immunological markers were seldom used to study plant variation, perhaps because few suitable proteins had been identified. During the past few years, however, monoclonal antibodies to seed proteins of several cereal species have been produced. Reactivity of these antibodies with antigenic proteins can be evaluated by enzyme-linked immunosorbent assays (ELISAs). Monoclonal antibodies have been used to identify cDNA library clones and describe variation between crop species and cultivars.

Other marker assays measure the rate of protein migration through a starch, polyacrylamide, or cellulose acetate gel in response to an electrical current. The migration rate is related to protein size, shape, isoelectric point and/or ionic charge.

In these assays, protein extracts from entries in an array are loaded in adjacent gel lanes and electrophoresed. The gel is then treated with a histochemical stain that reacts with a specific marker protein. Variation between the entries is measured by the positions of their stained proteins on the gel.

Many of the proteins used as markers are isozymes – functionally similar forms of an enzyme, encoded by different loci or different alleles at a locus. In general all alleles at a locus are expressed, so allelic isozymes can be analysed as codominant variants.

Isozymes are frequently used to describe variation between and within plant populations, species, and genera; for mapping and diagnostics; and in some cases for identification. When comparing higher taxonomic levels, the probability of genetically different electromorphs migrating to the same gel position is increased. For closely related taxa and populations, however, isozyme analysis is informative and continues to be refined.

Immunoelectrophoretic assays measure both protein migration rate in gels and the antibody specificity of proteins. Antigenic proteins are electrophoretically separated, then are detected by their reactions with antibodies. This technique is valuable for diagnostics, mapping, and describing variation between plant cultivars, species, and higher taxa.

Nucleic Acid Marker Assays: Total Genomic DNA and Chromosome Markers

Although proteins are useful as genetic markers, the phenotype each protein describes is determined by the genotype plus the type of tissue sampled, plant phenological stage, environment, and post-translational processing. In addition, the percentage of a genome sampled by protein markers is limited: these markers only describe coding sequences, and not all proteins can be separated and detected by established methods.

Some questions are best addressed by markers assumed to be selectively neutral, but this assumption is in question for many proteins. Finally, some proteins are unique to particular taxonomic groups, and thus are suitable as markers only in a limited range of taxa.

In contrast, nucleic acids are present in all organisms and are a common genetic currency for comparing virtually any taxa. Nucleic acid markers describe genotypes, not phenotypes, and can sample both coding and Non-coding regions of a genome. DNA methylation and secondary structure can cause artefactual variation in some assays, disrupting the direct relationship between marker and genotype. However, these artefacts can often be detected and eliminated.

The first nucleic acid marker assays measured properties of total genomic DNA. When centrifuged in a CsCl density gradient, DNA separates into bands according to GC content. Because highly repetitive (satellite) sequences differ in GC content from the rest of the genome, these gradients can be used to isolate and quantify satellite DNA. This method has been utilized to compare satellite DNA content between plant species, and is still used to detect highly repetitive DNA.

Another technique that separates repetitive and low-copy DNA relies on DNA reassociation kinetics. The two complementary strands of a DNA molecule denature when heated and reassociate into duplexes when cooled. At a standard temperature and salt concentration, reassociation rate depends on the percentage of strands with similar sequences. Repetitive sequences reassociate rapidly and at high temperatures, compared to single-copy sequences. Thus highly repetitive, middle repetitive, and single-copy fractions of a genome can be isolated by fragmenting, heating, and cooling double-stranded DNA. Because reassociation rate is affected by fragment length and the pattern of sequence repetition, the type and dispersion of repeats can be measured by varying the lengths of the fragments analysed. This method has been used to characterize and compare repeat sequences between plant species and hybrids.

DNA–DNA hybridization employs the same principles, but with single-copy sequences, to compare entries in an array. When single-stranded DNA from two entries is combined, their overall sequence similarity is related to the temperature at which 50per cent of the strands form heteroduplexes. This *T*m is compared to the homoduplex *T*m for each entry; D*T*m measures the genetic distance between entries. DNA–DNA hybridization is used to examine variation between species and higher taxa, usually with vertebrates and less often with plants and bacteria. Variation in DNA content of nuclei and chromosomes can be measured by flow cytometry. Particles are isolated, suspended, stained with a DNA-binding fluorochrome, and passed through a cytometer. The fluorescent signal intensity from each particle indicates its DNA content. Flow cytometry was first utilized to analyse mammalian genomes, but has since been used to compare nuclear DNA content and ploidy level between plant families, genera, species, and ecotypes and to evaluate wide hybrids. Because chromosomes can be identified by their signal intensity, flow cytometry can be used to generate high-resolution karyotypes and sort chromosomes for mapping.

Karyotypes of chromosomes prepared in mitotic metaphase squashes have long been used for diagnostics, mapping, and describing genetic variation between plant species and hybrids. Entries in an array can be characterized by their chromosome number and morphology. In addition, chromosome spread preparations can be treated with stains specific to AT-rich regions, heterochromatin, or nucleolar organizing regions. The stain-specific banding patterns of individual chromosomes are then used as markers to characterize and compare entries.

Chromosome banding is informative for mapping studies and describing variation between genebank accessions and cultivars. Banding is often used in combination with flow cytometry or other techniques. For chromosomes that are very small or lack distinctive bands, cytogenetic techniques may have limited use. As high-resolution microscopes and more sensitive stains are developed, however, the value of these assays continues to increase.

Nucleic Acid Marker Assays: Fragment Markers

Nucleic acid fragments are widely used as genetic markers. As mentioned above, CsCl density gradients and DNA reassociation techniques can detect and roughly quantify satellite DNA sequences. Yet these methods do not measure discrete character differences or variation in the sequences themselves. In addition, differences in GC content between genomes can significantly affect the results. In contrast, fragment markers provide high-resolution qualitative information about sequence variation. Either DNA or RNA can be evaluated, and minimal amounts are required. Numerous fragment marker assays are available; all involve detecting fragments that differ in presence, size, or quantity between entries in an array.

Fragments that are produced, electrophoresed, then detected: In many assays, fragments are produced, separated by agarose or polyacrylamide gel electrophoresis, then detected. The length and conformation of a fragment determine its migration rate in a gel. Several different approaches are taken for detecting fragments. In some assays with purified sequences, all fragments are detected and visualized.

The fragments are either labelled before or stained after electrophoresis. Ethidium bromide, silver, fluorescent, and radioactive stains and labels are commonly used. In other assays, specific target fragments are detected and visualized by transfer hybridization. All fragments in an assay are electrophoresed, denatured if double-stranded, then transferred from the gel to a nylon or nitrocellulose membrane. A labelled single-stranded probe complementing the target sequence is hybridized to the membrane and anneals to the target fragments. The extent of mismatch hybridization varies with salt and formamide concentration and the temperature during hybridization.

A variety of probes can be used: Synthesized oligonucleotides, uncharacterized clones from genomic DNA libraries, fragments isolated from other gels, RNA sequences, or total genomic DNA. The probe may be from the same species as the membrane-bound DNA or from a different species. Until recently most probes were radioactively labelled, but Non-radioactive colorimetric, fluorescent, and chemiluminescent labels are increasing in popularit).

Fragments produced by restriction digest: Restriction site analysis is widely used in plant genetics. The principles, protocols and data interpretation have been discussed in detail. Restriction fragments are generated by treating double-stranded DNA with restriction endonucleases (REs), which are enzymes produced by bacteria. Each RE cleaves double-stranded DNA at a specific sequence, usually four to six nucleotides in length. Several hundred REs have been characterized; they vary in recognition sequence and ability to digest methylated DNA. DNA sequence variation between the entries in an array is detected as variation in the number and/or lengths of restriction fragments. This variation can be caused by point mutations at RE recognition sites or by the insertion, deletion or rearrangement of sequences at or between RE sites.

When single- or low-copy fragments are evaluated, variations in fragment length are referred to as restriction fragment length polymorphisms (RFLPs). In some RFLP assays of highly purified DNA sequences, all of the fragments are visualized by labelling them before or staining them after electrophoresis. In most RFLP assays, however, target fragments are detected by hybridized probes. For each entry in an array, the presence or absence of each individual fragment is recorded.

These fragment data are used to infer the presence or absence of each restriction site. For diploid genomes, entries heterozygous for the presence of a restriction site can be distinguished from those that are homozygous. Although few RFLP loci are examined on each gel, every locus can be genetically defined. RFLP analysis can be expensive, but costs per trial can be reduced by probing each membrane several times with different probes. Single- or low-copy RFLPs are used extensively in diagnostics, for mapping, and to verify interspecific hybridization and study genetic relationships and structure. Probes that hybridize across species are especially useful for comparative mapping.

In most restriction site analyses of repetitive DNA, target fragments are detected by probing with a known repeat sequence. The fragments, not the genetically defined sites, are usually the characters analysed.

Entries that are homozygous for the presence of a fragment cannot be distinguished from heterozygotes. These assays provide information about variation at many loci simultaneously, but reveal little about the type of sequence variation. Restriction site analysis is often used to study interspersed repeats and satellites. It was utilized in the first assays of hypervariable loci, and is presently the most common method in forensics for evaluating minisatellite VNTR loci. Multilocus restriction fragment profiling of plant genomes is useful for identifying clonal cultivars and studying breeding systems. In some studies, fragment length polymorphism can be interpreted as variation of codominant alleles at a locus. However, because inheritance of markers in multilocus profiles is seldom clearly defined, their use in phylogenetic and gene flow studies may be limited.

Restriction site analysis can generate highly reproducible results, if all steps of the process are carefully controlled. With a variety of probe– enzyme combinations available, assays of the nuclear genome can be used to address many questions about genetic variation at or below the species level. As in isozyme analysis, the probability of genetically different restriction fragments co-migrating on a gel is increased when higher taxonomic levels are compared. Because chloroplast genomes evolve slowly, chloroplast restriction site assays are often used to study variation between higher taxa. In some species, both chloroplast and nuclear genomes are biparentally inherited. In many plant species, however, chloroplasts are either maternally or paternally inherited. For these species, chloroplast restriction site analysis is very informative for studies of gene flow and interspecific hybridization, especially when used in combination with assays of biparentally inherited nuclear markers.

Fragments produced by the polymerase chain reaction: Fragment analysis began in the 1970s with the introduction of RFLPs, and grew exponentially a decade later when the polymerase chain reaction (PCR) was developed. PCR uses the principles of DNA reassociation and the action of DNA polymerase to amplify nucleic acid fragments *in vitro*.

Each PCR reaction solution contains a double-stranded DNA template, short single-stranded oligonucleotide primers, thermostable DNA polymerase, enzyme cofactors, and the four deoxynucleotides (dNTPs: dATP, dCTP, dGTP, dTTP).

The template is denatured by heating; the temperature is lowered, and the primers anneal to complementary sequences on the template. The solution is heated again, and the polymerase adds dNTPs to the 3' end of each annealed primer. Double-stranded fragments are produced and serve as templates for the next cycle, which generates fragments with primer sequences at each end. During subsequent cycles, these fragments are amplified at a geometric rate.

With automated thermocyclers and cheaper, improved reagents, PCR is fast becoming more reproducible and affordable. A wide range of known and unknown sequences can be amplified from fresh or ancient DNA and from RNA, and little template is needed; hence PCR is applicable to any taxon. However, contamination is a concern, since contaminant DNA in a reaction solution can function as a template. The specificity and reproducibility of PCR depend on several factors: the concentration and quality of the reaction ingredients; the design and GC content of the primers; and the temperature, duration and number of cycles. However, measures can be taken to minimize and detect contamination and optimize cycling protocols. The disadvantages of PCR are greatly outweighed by the tremendous opportunities it provides. When amplified fragments are used for high-resolution mapping, diagnostics or forensics, they are generally detected by probes. When used to describe variation between and within species, amplified fragments may be detected by labelling primers or dNTPs in the reaction solution, labelling PCR products, or staining gels after electrophoresis.

Numerous PCR-based marker assays have been developed and used in plant genetics. They can be categorized by whether target sequences are known prior to amplification, whether the primer sequences are designed or are arbitrary, the number of primers, and the size range of amplified products.

In the first PCR assays the target loci were known single- copy sequences, amplified by two primers designed to complement the regions flanking the loci. Primer pairs have since been designed for numerous single-copy loci in plant and animal nuclear and organelle genomes.

These loci are very useful as markers for mapping, diagnostics, phylogenetic analyses, and studying hybridization.

In most genome mapping projects, cloned DNA sequences present at only one site in a genome are used as physical mapping landmarks. Primer pairs have been designed to amplify many of these landmarks as sequence tagged sites (STSs). This strategy is used extensively in the Human Genome Project (HGP) and other mapping programmes.

In assays of these designed-primer PCR loci, the character evaluated may be the fragment itself, with presence dominant over absence. Often, however, inheritance can be defined. For loci in diploid genomes, differences in fragment length can then be analysed as codominant alleles. Both interspersed and tandemly repeated sequences can be amplified by designed-primer PCR.

When several interspersed repeat loci are amplified in one sample, each locus is evaluated separately. When tandem repeat loci are amplified, length polymorphism at a locus is assumed to reflect variation in the number of tandem repeat units.

Amplification of human minisatellites with designed primer pairs was soon followed by amplification of plant minisatellites and SSRs. Amplified plant SSRs are informative for mapping, genotyping, diagnostics, and population studies. SSR alleles may differ little in length, and discriminating them can be difficult. New electrophoresis, labelling and imaging technologies are making detection easier.

Development of designed-primer PCR markers involves several steps. Target sequences are identified and isolated, usually from a genomic DNA library; the loci and flanking regions are sequenced; and primer pairs are designed and synthesized. Currently these tasks are expensive, and not all laboratories have the equipment required. Once primers are designed and synthesized, however, the costs for each analysis are not high. The initial expenses of marker development may be reduced by identifying the desired sequences from nucleic acid databases, *e.g.* Genbank or EMBL. However, this is only applicable to species well represented in the databases. As another cost-cutting measure, some primers designed to amplify loci in one species can be used to amplify loci in related taxa. Primers that amplify across taxa are also useful for phylogenetic analyses and studying genome evolution.

For development of designed-primer PCR markers, both target and flanking region sequences must be known. Soon after PCR was introduced, however, its ability to amplify unknown sequences was realized. In inter-repeat PCR, primers that complement interspersed repeats or other known sequences are used to amplify unknown sequences between the repeats. 'Universal' primers have been designed to complement conserved coding sequences in mitochondrial and chloroplast genomes; these primers can be used to amplify the Non-coding regions between genes.

Other marker assays require no prior sequence information, for either the target region or the flanking regions. These assays – randomly amplified polymorphic DNA analysis (RAPD), arbitrarily primed PCR (AP-PCR; Welsh and McClelland, 1990), and DNA amplification fingerprinting (DAF) – use one short primer that has an arbitrary sequence. This primer anneals to the template at complementary sequences in both 'sense' and 'antisense' orientation, and the regions between primers in opposite orientation are amplified. The number and length of fragments amplified by one primer can vary; short primers generally produce a large number of small fragments. In single-primer PCR, entries that are homozygous for the presence of a fragment cannot be distinguished from heterozygotes; thus fragment presence is generally dominant over absence. Many loci can be evaluated simultaneously, although little genetic information about each locus is provided. Arbitrary PCR is particularly valuable when no sequence information is available for the taxa being studied. It has been used extensively to examine variation between and within plant species, identify genebank accessions and cultivars, and construct genetic maps.

Many PCR assays are variations or combinations of designed-primer PCR and arbitrary PCR. Most were developed to increase the reproducibility of results and/or to better discriminate between closely related genotypes. In arbitrary PCR, reproducibility and the extent of mismatch primer annealing are highly dependent on the cycling parameters and concentration of reaction components. One solution to this problem is to identify polymorphic fragments in arbitrary PCR assays and then isolate, clone, sequence and design primers for these fragments. This process generates new designed-primer PCR loci but involves many steps. Other assays with increased reproducibility combine features of single-primer and interrepeat PCR: unknown sequences are amplified by one primer that complements an SSR, its flanking region, or both. Finally, 'anchored' PCR uses one designed and one arbitrary primer to amplify unknown sequences.For genotyping, diagnostics, and assessing variation between close relatives, PCR-based marker assays with high resolution are needed. One such variation is nested PCR. In this technique, one amplification reaction with two designed primers is followed by a second reaction. The first PCR product is the template for the second reaction, which uses two designed primers internal to the first set. Although the possibility of contamination is high, nested PCR is valuable for extremely sensitive genotyping and diagnostics. At present, it is used most often in medical research.

Resolution of arbitrary PCR can be increased by adding stable mini-hairpin sequences to the 5' ends of short primers. This technique greatly improved discrimination between cultivars of centipedegrass [*Eremochloa ophiuroides* (Munro) Hackel]. Other marker assays with improved resolution combine principles of PCR and restriction site analysis. RE digestion of PCR templates before they are amplified was first suggested by Mullis and Faloona (1987). Arbitrary PCR with cleaved templates has since been used to map plant genomes and identify genotypes. Amplified fragment length polymorphism (AFLP) analysis is a variation of cleaved-template PCR. After DNA is digested, a short adaptor sequence is ligated to each end of the restriction fragments. The fragments are then used as templates for single-primer PCR. The primer contains a constant sequence (the adaptor and RE recognition site sequences) plus selective bases that determine the specificity of primer annealing. Both reproducibility and resolution are increased by using AFLPs.

Resolution can also be increased by digesting PCR products before electrophoresis. Digested products of both designedprimer PCR and arbitrary PCR have been used to measure variation between genebank accessions, refine RFLP maps, and study interspecific gene flow. Resolution of both PCR and restriction site assays can be improved by denaturing double-stranded DNA fragments before they are electrophoresed. Due to internal point mutations or differences in conformation or GC content, the two single strands from one fragment may have different migration rates on polyacrylamide gels. Both denaturing gradient gel electrophoresis (DGGE) and single-stranded conformation polymorphism (SSCP) analysis can detect differences between closely related genotypes.

Fragments detected directly by probe hybridization. In some marker assays the target sequences are not separated by electrophoresis, but are detected directly by probe hybridization. The types of probes used are the same as in electrophoretic assays, and can be radioactively or Non-radioactively labelled.

In dot and slot blot hybridization, single-stranded DNA fragments, genomic DNA, or RNA is immobilized on a nylon or nitrocellulose membrane. Labelled probes complementing the target sequence are hybridized to the membrane blot. All loci in the blot are sampled simultaneously; abundance of the target sequence is measured by the intensity of the hybridization signal. Many blots on one membrane can be evaluated at the same time, and the procedure is relatively simple. Because the presence of Non-target sequences in the blot can affect hybridization, precise measurment of target sequence abundance in genomic DNA blots can be difficult. When the blots are carefully prepared and hybridization conditions are controlled, however, these assays can generate valuable information for diagnostic and relationship studies. Dot and slot blots have been used to detect specific alleles of genes and RAPD loci, estimate the copy number of satellites and interspersed repeats, and compare gene expression between organs of a plant.In another technique, *in situ* hybridization (ISH), probes are hybridized to the same chromosome spread preparations used for karyotypes and banding pattern assays. ISH is used to detect the presence and chromosomal distribution of target sequences. While banding assays are only appropriate with chromosomes that contain specific sequences, ISH can detect a wide range of sequences. Radioactive and Non-radioactive probes are used, but the popularity of fluorescent *in situ* hybridization (FISH) is rapidly increasing. ISH has many applications for genome analysis. These include 'painting' the genome of one species onto chromosomes of another, to study species evolution; painting a genome with one of its own chromosomes, to study chromosome evolution; and characterizing the chromosomal distribution of interspersed repeats. ISH is often used in combination with other marker assays.

In wide crosses and somatic hybrids, ISH with species-specific interspersed repeat probes can indicate the abundance and chromosomal locations of transferred sequences; restriction site analysis can provide information about structure and rearrangement of the sequences. Physical mapping with ISH can complement RFLP genetic mapping studies. Aledo *et al.* used slot blots to estimate the copy number of an interspersed repeat in the maize (*Zea mays* L.) genome, and used FISH to determine the chromosomal distribution of the repeat. With both ISH and heterochromatin-specific staining on the same chromosome preparation, Harrison and Heslop-Harrison (1995) confirmed that two *Brassica* satellite families are located near centromeres. By varying the hybridization stringency, differences between satellite sequences on different chromosomes were revealed.

ISH can be time consuming, and results can be affected by the specific methods used for chromosome preparation, denaturation, and probe hybridization and detection. However, new protocols are increasing the reproducibility of ISH. Fukui *et al.* (1994) used a thermocycler to control temperatures during denaturation, and developed accurate imaging methods for detecting and analysing fluorescent signals. Because of such innovations, the value of ISH as a molecular tool will most likely continue to increase.

Nucleic Acid Marker Assays: Nucleotide Sequencing

Nucleic acid variation can be described most directly by using the nucleotides themselves as discrete characters, each with four possible states. Because nucleotides are the basic units of genetic information in all organisms, sequencing can be used to detect very low levels of variation. Nuclear and organelle genomes of virtually any taxon can be sequenced; minimal amounts of DNA or RNA are needed, and fresh or preserved DNA can be used. With most marker assays, entries in a trial are compared to each other; results are not easily compared across trials that have no standard entries in common. In contrast, sequences from one study can be compared with sequences from any other study. In addition, sequencing can reveal whether variation is due to nucleotide substitution or sequence rearrangement.

Two sequencing methods have been used since the late 1970s, and are described in detail by Hillis *et al.* (1990). In both methods, labelled fragments are generated from DNA or RNA that has been isolated and purified; the fragments are electrophoresed; and the sequence is 'read' from the fragment patterns. Maxam–Gilbert (chemical) sequencing is used to sequence double-stranded DNA. DNA is endlabelled, then divided into four reaction solutions.

Each solution contains one of four reagents, each of which cleaves DNA at specific bases. The sequence of the DNA determines the lengths of the fragments produced. *In vitro* (dideoxy chain termination) sequencing is applicable to both DNA and RNA. The single-stranded template is divided into four reaction solutions. Each solution also contains the following: a labelled oligonucleotide primer that complements one end of the template; the four dNTPs; and one of four dideoxynucleotides, each of which terminates extension at one of the four bases. In each reaction, the primer is extended along the template. As in chemical sequencing, the template sequence determines the lengths of the fragments produced. Chemical sequencing is preferred for some genomic regions with complex secondary structure; in general, however, the chain-termination method is more commonly used. Since it has been coupled with PCR, combined with fluorescent labelling and automation, this technique is rapidly becoming the method of choice. Sequencing can be used to address a wide range of plant genetic resource questions about virtually any taxa, but involves considerable investment of money, labour and time. Automation and large-scale production of cheaper reagents have made sequencing more cost effective, but few laboratories can afford to sequence large genome segments from multiple entries. As a result, in many studies only one locus or a few loci are sequenced for a limited test array. While many adjacent nucleotides may be evaluated, together they sample only one locus. Thus sequencing is usually not practical for mapping, multilocus forensics, population genetics, or analyses of intraspecific variation. It is often used for high-resolution diagnostics, studying specific genes, and describing variation between genera and higher taxa. In phylogenetic studies, however, inferring genome evolution from sequence variation at a single locus should be done with caution if at all.

Many regions of plant nuclear and organelle genomes have been sequenced and used as markers. These sequences vary greatly in evolutionary rate and the taxonomic levels at which they are employed.

Chloroplast genomes evolve slowly; sequences of the entire genome or its *rbc*L gene are informative as markers for describing variation between genera and higher taxa. Nuclear genes that have been sequenced and used as markers include the phytochrome, heat shock protein, and nuclear rRNA gene families. In general, the rRNA coding regions are highly conserved; internal transcribed spacers are more variable, and intergenic spacers are highly polymorphic. Finally, sequencing is often used – in combination with other marker assays – to characterize tandem repeat loci. While sequencing may be the ultimate molecular tool, caution should be taken to avoid several potential problems. Because each nucleotide is a separate character, sequence fidelity is crucial. Thus nucleic acids must be carefully isolated and purified before they are sequenced.

In addition, PCR contamination and amplification errors have more serious effects for sequencing than for fragment analysis. Amplification errors can be minimized by optimizing PCR conditions and using polymerases with proofreading capability.

Mistakes during gel reading (manual or automated) must also be considered and minimized. Subjectivity is a real concern when aligning sequences before they are compared. Alignment is complicated and requires assumptions about expected amounts of base substitution and insertion/deletion. Despite these concerns, sequencing remains the tool of choice for many studies.

FUTURE DIMENSIONS OF GENETIC ANALYSIS

The goals and expectations for analysing plant genetic variation parallel those established across many other fields of biological research, from agriculture, ecology, and evolution to the medical sciences. In all of these fields, future genetic marker assays must incorporate methods to detect, describe, interpret, and store DNA sequence information. Molecular tools of the future are expected to be user friendly, accurate, precise, high throughput, low cost, and potentially automated.

DNA sequence information is the foundation for developing and applying genetic markers to questions of biological variation, whether *in situ* or *ex situ*. Researchers who develop and use sequence-based marker assays for diagnostic, forensic and relationship studies will continue to benefit greatly from information and technologies generated by the international Human Genome Project (HGP). Since its beginning, the HGP has endeavoured to develop genetic and physical maps and determine DNA sequences for the genomes of humans and several model organisms. As part of this effort, genetic analysis methods have been improved significantly. Notable achievements – now taken for granted by many

– include new types of genetic markers (particularly those based on PCR), more efficient cloning vectors, the introduction of STS markers, and improved technology and automation for DNA sequencing. Underlying this progress is recognition that the successful development of new technologies for genetic research has been, and will continue to be, critical to many future scientific breakthroughs.

In addition to the HGP, other complementary projects also support initiatives for the development of new molecular marker assays. One such project is the Advanced Technology Programme (ATP) of the US National Institute of Standards and Technology. Among the ATP's goals is working with industry to deliver DNA diagnostic tools to a variety of users, at onetenth to one-hundredth of the current price. Another goal is helping industry make similar reductions in DNA sequencing costs and make DNA sequencing apparatus available at about one-third of the current price.

Many biological questions once considered recalcitrant due to the number and cost of the assays required will be reconsidered in the light of new technological developments. The following section includes examples of progress that can be readily applied to the description of plant genetic variation. Also noted are limitations that are unique to plant-based studies, and are not presently addressed by most genetic analysis programmes.

GUIDELINES FOR SELECTING MARKER ASSAYS

Choosing appropriate marker assays can be challenging, but several considerations can make the task easier. Important issues are: (i) what question is being asked? (ii) what level of resolution is required? (ii) how can the results be related to characteristics of the taxa being studied? and (iv) are sufficient resources available in terms of personnel, equipment, finances, and time?. Questions of identity are best addressed with high-resolution assays and large numbers of codominant marker loci that are evaluated sequentially.

Electrophoretic assays of isozymes and nucleic acid fragments meet most of these criteria. Genetically defined isozymes, amplified SSR loci, and single-copy RFLPs are very informative markers for identification. Due to practical constraints, however, multilocus assays of restriction fragment and PCR markers (VNTRs, RAPDs and AFLPs) are often used. Dot blots, sequencing and monoclonal antibody assays are also appropriate, but the costs of analysing many loci per assay may be prohibitive. The most appropriate assays for diagnostic questions have high levels of resolution and use genetically defined, codominant markers. Because these questions can often be addressed with one or few loci, useful methods include electrophoretic assays of isozymes, single-copy RFLPs, or SSRs; dot or slot blot assays; and sequencing.

For some questions, monoclonal antibody assays and ISH are also appropriate. Mapping studies often use a combination of tools: isozyme or DNA fragment electrophoresis for genetic maps, and chromosome banding, flow cytometry, ISH, and STS markers for physical mapping.For questions of relationship and structure, important issues include the type of question asked and the range and levels of taxa evaluated. Because evolutionary questions must be answered with qualitative data, the assays used for diagnostics and identification are appropriate. Questions of similarity can be addressed with these qualitative marker assays or with quantitative assays (microcomplement fixation, flow cytometry, CsCl density gradients, and DNA–DNA hybridization). The latter two techniques have low levels of resolution, and are generally used to compare genera and higher taxa. Interspecific and intraspecific variation is more appropriately described with isozyme and DNA fragment markers, which have higher levels of resolution.

In electrophoretic assays of isozymes and DNA fragments, co-migration of genetically dissimilar electromorphs or fragments is likely when higher taxa are compared. If resolution is too high, however, the assays may be numerically precise but contain random 'evolutionary noise' that masks meaningful variation. Researchers can often decide whether a particular marker assay is suitable for a given study by reviewing the results of previous studies using that technique.

Several characteristics of the taxa being studied are important when selecting a marker and marker assay. One of these characteristics is the mating system. Many apomictic and self-pollinated plant species contain little variation, so techniques with high resolution may be required. When deciding whether to use a chloroplast genome marker, information about chloroplast inheritance is critical. For many marker assays, polyploidy can be a concern. When polyploid taxa are evaluated, the number of loci sampled, the problem of dissimilar fragments co-migrating, and the level of data complexity are often increased. This complexity may be reduced by using genetically defined isozymes, chloroplast genes, genes that evolve slowly, or single-copy markers – rather than RAPDs or multilocus VNTR profiles.

Practical considerations are often the most important. The amount and quality of proteins or nucleic acids required can differ greatly between marker assays. Sequencing and PCR-based assays require little template and can be used to analyse ancient DNA samples, but are very sensitive to contamination. DNA–DNA hybridization requires large amounts of DNA, but small amounts of contaminant DNA are less of a problem. For some species, fresh tissue is required for isozyme electrophoresis. For some species and marker techniques, the presence of secondary compounds can complicate storage and extraction of DNA or protein, and can interfere with marker detection.

Marker assays also vary in the amount of setup work required and the extent of sequence information needed before conducting the assay. Karyotyping, ISH, and RFLP assays involve preparing chromosome spreads and/or probes. In contrast, some PCR-based assays are streamlined for high throughput and require little setup work. Sequence information is needed for probe hybridization and designed-primer PCR assays, but not for arbitrary PCR. Finally, the resources available must be considered. Minimal equipment is needed for protein electrophoresis, but other techniques require thermocyclers, fluorescent microscopes, cytometers and radioactive labelling.

Choices between newly developed and well-documented techniques should be made in light of the expertise and personnel available. Other considerations are reproducibility, number of entries and loci sampled per analysis, and cost per unit of information. Comparative studies can make selection of a marker assay easier and reveal where costs can be reduced. Costs and technologies are constantly changing, however, and the comparisons themselves may be expensive and time consuming; their value must be weighed against the resources they require and the necessity of continual updates. When comparative studies are not feasible, thoughtful consideration of the issues described here may be helpful.

CONCLUSION

The conservation of plant genetic resources has, in recent years, attracted growing public and scientific interest and political support. There is an increasing awareness of the relevance of biological diversity and its conservation to the health of the biosphere. Particularly urgent is the need to raise agricultural output to meet the basic nutritional needs of increasing populations.

The dilemma is that the required increase in food production must be obtained through 'sustainable' forms of agriculture which are less dependent on the use of modern high-yielding varieties bred for intensive, high-input systems. It seems unlikely that increased food production on this scale will come from increasing the area under cultivation. Rather, major solutions will depend upon innovations which exploit sophisticated methods to manipulate plant genomes and the totality of natural genetic variation.

The area of disease resistance will serve to illustrate this argument. Many authors have emphasized the importance of innovation in the area of crop protection since crop losses due to pests and disease may account for between 20 and 40per cent of productivity worldwide.

Indeed, over 50per cent of the research and development (RandD) spent in the plant breeding/seed industry is focused on the identification and incorporation of resistance traits. In a situation where there is some evidence that the pool of available variation for disease resistance genes within commercial breeding lines is becoming limited (*e.g.* for yellow rust in cereals) the search is on for novel sources of resistance.

There are now many examples of biotechnology being put to the service of pest control, through the incorporation into plants of transgenes for toxins, antifeedants, enzyme inhibitors, etc. But, as Woolhouse (1992) has pointed out, such is the long coevolutionary history of pests and pathogens with their hosts that present strategies for interfering with these relationships represent only a small proportion of the likely defence mechanisms used by plants in resisting foreign organisms. This enormous diversity of mechanisms represents an arsenal of ammunition which can be used in the fight to minimize disease and predation. Are there any hard facts which support these assertions concerning the value of genetic resources to agricultural productivity? Can we put a cost on the loss of diversity? The sources of variation used by the plant breeder range from current breeding lines to wild species and the products of direct gene manipulation. In a recent analysis of the plant breeding/seed industry, Swanson (1996) showed that, over a five-year period, 6.5per cent of all genetic research within this industry, resulting in a marketed innovation, was concerned with germplasm from wild species and landraces.

This compares with only 2.2per cent of new germplasm arising from technological approaches of induced mutation. The vast majority of germplasm successfully used by the industry still arises from 'conventional' sources, *i.e.* the heavily exploited commercial cultivars, which at first sight might be taken to suggest that wild resources are relatively unimportant in RandD.

In fact Swanson (1996) argues just the opposite, suggesting that the stock of existing commercial varieties should be seen as the information base from which bioindustries develop innovations, whereas the sources of new diversity (wild species, landraces, induced mutation) should be seen as supplying increments to the information base. In other words, RandD requires an annual injection of 'new' genetic material from natural sources, estimated as amounting to approximately 7per cent of the stock of material currently within the system.

Biodiversity is valuable to industries other than those that seek novel genes for crop plant improvement. Biodiversity has traditionally provided a source of compounds to the pharmaceutical, foodstuffs, crop protection and other industries and here it is possible to give more precise evaluations of its importance. As outlined in Chapter of this volume, drug sales based on natural products from plants were estimated at $US43 billion in 1985 alone and the value of yet undiscovered pharmaceuticals in tropical rainforests is estimated to be as much as $US147 billion to society as a whole. Vast markets exist for the replacement of synthetic biocides, many of which have associated toxicological and environmental problems, by 'natural' compounds, which may have more specific modes of action and fewer side-effects.

Even greater may be the use of genetic resources in more fundamental research where their value in the longer term may be unknown or impossible to estimate accurately at the present time. Diverse genetic material is needed and now regularly utilized to create mapping populations and to study inheritance of a wide range of traits which may or may not be of immediate practical value. Very large numbers of germplasm samples are regularly screened and evaluated for different characteristics so that the underlying molecular biology can be investigated. The intention here is that a greater understanding of fundamental biology will give rise to wealth creation in the future.

THOUGHTS ON THE FUTURE

In the HGP, technological improvements unanticipated in 1990 have already changed the scope of the research and allowed for more ambitious approaches and goals. In the plant kingdom as well, progressive visions of DNA analysis will most likely change the ways in which problems related to describing genetic variation are perceived and resolved. Beyond the marker assays discussed above, another approach to automated DNA diagnostics merits mention.

Sequencing by hybridization (SBH) uses a large number of short oligonucleotide probes, immobilized in an array on a small solid base. Single-stranded, fluorescently labelled DNA of an unknown sequence is hybridized to the probes on this 'DNA chip', and the sequence is determined from the hybridization pattern. SBH may be the ultimate application of 'clinical' sequencing to the description of genetic variation. Prior to widespread use of SBH, however, new approaches to synthesizing oligonucleotides, attaching them to solid surfaces, and fabricating the microchips will be required.

In order to better characterize plant genetic diversity and address genetic resource questions, plant scientists will need molecular marker assays that cost-effectively detect and describe DNA sequence variation over many areas of the genome (both coding and Non-coding), for many individuals in a population or taxon. This goal may seem daunting, but much progress is being made. As long as plant scientists are aware of and build on innovations in other fields, a stream of new tools and approaches will be available for the challenge.

Bibliography

A. Bandyopadhyay , K.V. Sundaram , M. Moni , M.M. Jha and P.S. Kundu: *Sustainable Agriculture*, Northern Book Centre, Delhi, 1996.

A. Rashid: *Introduction to Genetic Engineering of Crop Plants : Aims and Achievements*, I.K. International Publication, Delhi, 2009.

Anil Kumar Srivastava: *Genetic Engineering and Biotechnology*, Swastik Publication, Delhi, 2011.

B K Desai and B T Pujari: *Sustainable Agriculture : A Vision for Future*, New India Publication, Delhi, 2007.

C.P. Malik: *Genetic Engineering : A New Hope for Crop Production and Improvement*, Aavishkar Publication, Delhi, 2010.

E. Thro: *Genetic Engineering : Shaping the Material of Life*, Universities Press, Delhi, 2000.

Gyan Deep Singh: *Genetic Engineering of Plants*, Anmol Publication, Delhi, 2008.

Jane K. Setlow: *Genetic Engineering : Principles and Methods*, Springer, New York, 2010.

L. Yount: *Genetics and Genetic Engineering*, Universities Press, Delhi, 1999.

M. Follower: *Genetics and Genetic Engineering*, Rajat Publication, Delhi, 2003.

N.K. Yadava: *Molecular Biology and Genetic Engineering*, Manglam Publication, Delhi, 2009.

Nicholl: *Introduction to Genetic Engineering*, Cambridge University Press, New York, 1999.

P N Kalla; Anita Singh; S S Pareek; Shanti K Sharma and Hanuman Ram: *Sustainable Agriculture : Status and Prospects*, Agrotech Publishing Academy, Delhi, 2007.

Pete Shanks: *Human Genetic Engineering : A Guide for Activists, Skeptics, and the Very Perplexed*, Discovery Publishing House, Delhi, 2007.

Pratik Satya: *Genomics and Genetic Engineering*, New India Publication, Delhi, 2007.

Preeti Joshi: *Genetic Engineering and Its Applications*, Agrobios Publication, Delhi, 2003.

R. Arunachalam and R. Netaji Seetharaman: *Sustainable Agriculture : Indigenous Practice for Natural Resource Management*, Agrobios Publication, Delhi, 2004.

R.K. Behl, D.P. Singh, V.S. Tomar, M.S. Bhale, D. Khare and S.D. Upadhyaya: *Sustainable Agriculture for Food, Bio-Energy and Livelihood Security*, Agrobios Publication, Delhi, 2009.

Rabindra Narain and Surendra Naha: *Genetic Engineering in Plants*, Dominant Publication, Delhi, 2006.

Rajiv K. Sinha: *Sustainable Agriculture : Embarking on the Second Green Revolution with Technological Revival of Traditional Agriculture*, Surabhi Publications, Delhi, 2004.

Ramesh Umrani and C.K. Jain: *Sustainable Agriculture*, Oxford Book Company, Delhi, 2010.

S R Kulshrestha: *Genetic Engineering*, Sonali Publication, Delhi, 2007.

Setlow: *Genetic Engineering: Principles and Methods*, Springer, New York, 2001.

Sheela Srivastava: *Genetic Modified Crops and Genetic Engineering*, Manglam Publication, Delhi, 2011.

Smita Rastogi and Neelam Pathak: *Genetic Engineering*, Oxford University Press, Delhi, 2009.

Surendra Naha and Ravindra Narain: *Genetic Engineering*, Dominant Publication, Delhi, 2004.

Surendra Naha; S Banerjee and Nandan Hazare: *Molecular Biology and Genetic Engineering*, Dominant Publication, Delhi, 2009.

V Kumar Gera: *Genetic Engineering and Biotechnology*, Dhruv Publications, Delhi, 2006.

Vandana Shiva and Gitanjali Bedi: *Sustainable Agriculture and Food Security : The Impact of Globalisation*, Sage Publication, Delhi, 2002.

Wagdy A. Sawahel: *Plant Genetic Engineering From A to Z*, Daya Publication, Delhi, 1997.

Zamir K. Punja: *Fungal Disease Resistance in Plants : Biochemistry, Molecular Biology and Genetic Engineering*, International, Delhi, 2005.

Index